UNIVERSITY OF BRIDGEPORT
MAGNUS WAHLSTROM LIBRARY
126 PARK AVE.
BRIDGEPORT, CT 06601

Anne Kathleen Dougherty, MA

Herbal Voices
American Herbalism Through the Words of American Herbalists

*Pre-publication
REVIEWS,
COMMENTARIES,
EVALUATIONS . . .*

"This is an outstanding book. It is the first of its kind—a boundary-breaker between allopathic and herbal medicine. Anne Dougherty clearly illustrates how herbs can be used in the same way as allopathic drugs but goes far beyond that 'botanical' paradigm to describe the myriad possibilities for integrating herbalism into everyday life. She expresses the importance of realizing that the 'whole plant,' when picked and utilized with conscious appreciation and awareness, can have healing benefits not realized under the industrial paradigm. For her, the 'whole' means the whole plant, the whole person, and the whole of nature. Together, these can produce a stunning synergy of healing; separated, herbs can become no more than replacements for allopathic drugs.

Dougherty's appropriate use of systems theory for explicating individual and cultural relationships to the healing power of herbs extends her analysis beyond her subject, linking it to other such studies of alternative healing systems. Both as an anthropologist and as a frequent user of herbs for my own healing, I am captivated by this book, from beginning to end. I find it to be a work of information, holistic comprehension, and deep communication between the worlds of herbs, herbalists, and the wider society that is increasingly coming to appreciate and rely on their healing powers."

Robbie Davis-Floyd, PhD
Author, *Birth As an American Rite of Passage;* Co-author, *From Doctor to Healer: The Transformative Journey*
(from the Foreword)

More pre-publication
REVIEWS, COMMENTARIES, EVALUATIONS . . .

"Dougherty has written a remarkably readable account of herbalism in America. Her distinction among the types of herbalists is illuminating and dialogue-generating. Her discussion of spiritual herbalism is informative and captures the difference between indigenous cultures and their brand of herbalism and the modern botanical medicine that is filling health food stores. Her perspective on the economics of herbalism and its place in a capitalistic economy is also very insightful; she shows how the growth of herb sales has threatened, in some cases, the continued existence of certain herbs in the wild. Dougherty thoroughly reviews the ethics of wildcrafting and harvesting herbs and discusses options for long-term sustainability of herb production. Her interviews with American herbalists bring her discussion points to life through the unique stories of people representing the different herbalist persuasions found across North America.

I heartily recommend this book as good reading. It may become a classic in the sociology of herbalism and gives us a perspective on current conditions and how they arrived from the past, infiltrate our present, and may be projected into the future. It is entertaining and informative reading and will open the path for further dialogue about the pros and cons of the various approaches to herbalism."

Lewis Mehl-Madrona, MD, PhD
Coordinator for Integrative
Psychiatry and System Medicine,
University of Arizona
College of Medicine

"Anne Dougherty has created an invaluable reference for everyone interested in biomedicine, herbalism, and the healing arts. Through a series of in-depth interviews, she explores the ideals, realities, and challenges facing contemporary American herbalism as it evolves to meet the demands of time. Rather than being given one person's subjective view of a complex and evolving subject, we are instead invited into the exciting, multifaceted world of herbalism, and introduced to its potentials, its limitations, its incredible spiritual base, and its scientific merit. This book explores the heart and soul and brains of herbalism via those who know and understand it best: the practitioners of the green.

Perhaps the most important piece of the entire book is Anne's own vision and dream for American herbalism. She keeps that voice soft, but if we find the thread and follow it, we are left with a vision of a holistic paradigm where 'American herbalism is a sustainable, thriving healing art and science, a relevant part of modern life and health care.'"

Rosemary Gladstar
Founder, United Plant Savers
and The California School
of Herbal Studies; Author,
The Gladstar Story Book
Herbal Series and *The Gladstar Family Herbal*; Co-editor,
Planting the Future

NOTES FOR PROFESSIONAL LIBRARIANS AND LIBRARY USERS

This is an original book title published by The Haworth Integrative Healing Press®, an imprint of The Haworth Press, Inc. Unless otherwise noted in specific chapters with attribution, materials in this book have not been previously published elsewhere in any format or language.

CONSERVATION AND PRESERVATION NOTES

All books published by The Haworth Press, Inc. and its imprints are printed on certified pH neutral, acid-free book grade paper. This paper meets the minimum requirements of American National Standard for Information Sciences-Permanence of Paper for Printed Material, ANSI Z39.48-1984.

Herbal Voices
American Herbalism Through the Words of American Herbalists

THE HAWORTH INTEGRATIVE HEALING PRESS
Ethan Russo
Editor

The Last Sorcerer: Echoes of the Rainforest by Ethan Russo

Professionalism and Ethics in Complementary and Alternative Medicine by John Crellin and Fernando Ania

Cannabis and Cannabinoids: Pharmacology, Toxicology, and Therapeutic Potential by Franjo Grotenhermen and Ethan Russo

Modern Psychology and Ancient Wisdom: Psychological Healing Practices from the World's Religious Traditions edited by Sharon G. Mijares

Complementary and Alternative Medicine: Clinic Design by Robert A. Roush

Herbal Voices: American Herbalism Through the Words of American Herbalists by Anne K. Dougherty

The Healing Power of Chinese Herbs and Medicinal Recipes by Joseph P. Hou and Youyu Jin

Alternative Therapies in the Treatment of Brain Injury and Neurobehavioral Disorders: A Practical Guide edited by Gregory J. Murrey

Herbal Voices
American Herbalism Through the Words of American Herbalists

Anne Kathleen Dougherty, MA

The Haworth Integrative Healing Press®
An Imprint of The Haworth Press, Inc.
New York • London • Oxford

Published by

The Haworth Integrative Healing Press®, an imprint of The Haworth Press, Inc., 10 Alice Street, Binghamton, NY 13904-1580.

© 2005 by The Haworth Press, Inc. All rights reserved. No part of this work may be reproduced or utilized in any form or by any means, electronic or mechanical, including photocopying, microfilm, and recording, or by any information storage and retrieval system, without permission in writing from the publisher. Printed in the United States of America.

Cover design by Lora Wiggins.

Library of Congress Cataloging-in-Publication Data

Dougherty, Anne Kathleen.
    Herbal voices : American herbalism through the words of American herbalists / Anne Kathleen Dougherty.
        p. ; cm.
    Includes bibliographical references.
    ISBN 0-7890-2203-6 (hard : alk. paper) — ISBN 0-7890-2204-4 (soft : alk. paper)
    1. Herbalists—United States—Interviews.
    [DNLM: 1. Phytotherapy—United States—Interview. WZ 112 D732h 2004] I. Title.

RM666.H33D68 2004
615'.321'092273—dc22

2003026561

To all those who continuously push the bounds
of contemporary thinking—
who have the ability to know the Beyond
and the strength to bring it into the Now

ABOUT THE AUTHOR

Anne Dougherty is an herbalist, a health educator, and a freelance writer based in the Green Mountains of Vermont. She is also an Emergency Medical Technician active both in the ambulance and in the emergency room. In 1997, she and her partner founded Grian Herbs, an herbal products company specializing in organically grown and ethically wildcrafted alcohol extracts of bioregional botanicals. Ms. Dougherty earned her master's degree in Health Arts and Sciences from Goddard College where she pursued a groundbreaking, interdisciplinary study of the culture of American herbalism. Her particular interest lies in the overlapping of the herbal and biomedical communities, from which emerges newly forged identities.

CONTENTS

Foreword   xv
   Robbie Davis-Floyd

Preface and Acknowledgments   xxv

PART I: THE WEAVING

Chapter 1. The Dream: My Process and Methodology   3

Chapter 2. The Terrain: Consideration of Botanical Medicine and Holistic Herbalism   11
   Botanical Medicine   14
   Holistic Herbalism   20

Chapter 3. Recycling Science and Grasping the Ungraspable: Cartesian Thought and Folklore in Herbal Practice   33
   How Does Modern Biomedicine Enhance the Practice of Herbalism?   35
   What Does Herbalism Derive from Folklore?   39
   Observation and Change: The Link Between Science and Folklore   44

Chapter 4. Free-Form Goes Mainstream: The Debate over Licensure and Professionalization   49
   A Case for Licensure   50
   The Reservations   54

Chapter 5. The Ecology of Herbalism: The Role of the Herbal Community in the Endangered Plant Crisis   61
   The Role of Industry in the Endangered Plant Crisis   63
   What About Beyond Industry?   69
   Are Plants Really Endangered? An Herbalist's Perspective   78

**Chapter 6. Sinking Roots, Reaching Branches:
Spirituality and Tradition in Modern American
Herbalism**   **83**

The "Borrowing" of American Herbalism   89
Giving Back   90
The Herbalist and the Spirit: Unification   92

PART II: THE THREADS—INTERVIEWS WITH TWENTY AMERICAN HERBALISTS

Chapter 7. Laura Batcha   99

Green Mountain Herbs (GMH)   99
Running a Heart-Centered Business on the National Level   100
The Boom in Natural Products   102
Current Federal Regulations   103
Small Companies and Consumer Expectations   104
The Mission of GMH   105
Endangered Plants   106
Certifying Efficacy   106
Nature of Clinical Trials   108
Advice for People Wanting to Get into the Herb Business   109
Educating Consumers About Heart-Centered Herbal Products   110

Chapter 8. Richo Cech   113

Factors Contributing to At-Risk Plants   113
Criteria for Listing on the UpS At-Risk List   115
Role of Manufacturing in the Endangered-Plant Crisis   116
Rationale and Effect of Listing Goldenseal on CITES—Appendix II   117
Monopopularizing and the Limited Mainstream Materia Medica   119
Proactive Approach of UpS   120

Chapter 9. Ryan Drum   123

Herbalism and Its Renewed Popularity   123
Clinical and Lay Herbalists   124
Education in the Marketplace   126
Necessity of a Thorough Education for Practitioners   126

Clinician-Patient Relationship	129
Usefulness of Science and Folklore	132
Forging a Relationship with the Allopathic Community	133
Offering Herbs When Antibiotics Are Not Appropriate	134

Chapter 10. Daniel Gagnon — 137

Holistic Health	137
The Process Rather Than the Pill	139
More on Holistic Health	140
Is the American Public Healthier Today?	141
Helping the Mainstream Access Herbal Information	141
Medical Science and Its Interface with Herbalism	142
Science versus Folklore in Evaluating Herbs	142
Holistic versus Allopathic Herbalism	143
Mainstream Access to Holistic Health	143
Licensure for Herbalists	144
Education for Herbalists	145
Endangered Plants, Organic Growing of Herbs, and Diversifying the Mainstream Materia Medica	146

Chapter 11. Leslie Gardner — 149

Leaving Clinical Practice	149
Working in Deep Relationship with Plants	150
Commercialization of Herbs	151
Why the Herbal Renaissance Is Happening Now	152
Opening up the Mainstream to Holistic Herbalism	153
Steps an Herbalist Can Take to Preserve Herbalism	153
Ritual	154
Rediscovery of Ritual in America	154
What Makes Ritual Meaningful?	155
Rituals That Anyone Can Do	155
Plant Allies	156
Education for Herbalists	157
Benefits of Classroom Learning	158
Integrating Different Experiences of Herbs in Teaching	158
Science and Folklore in Herbalism	159
Herbalism in General	159

Chapter 12. Kate Gilday — 161
- At-Risk Medicinals — 161
- Commerce of At-Risk Medicinals — 162
- Educating Toward Alternative Plant Use — 163
- Ethical Wildcrafting — 163
- Evolution of the Plant-Human Relationship — 164
- Are Herbs Moving Deeper into the Woodlands? — 165
- Disconnection from Land — 166
- Domino Effect of At-Risk Medicinals — 168
- A "New" Way of Knowing Medicinal Plants — 168
- Herbs in the Mainstream — 169
- Making At-Risk Flower Essences — 170
- Hope for the American Herbal Community — 172

Chapter 13. Rosemary Gladstar — 173
- Experiential Learning of Herbs — 173
- Teaching About Herbs — 175
- Goals and Methods for Teaching Herb Classes — 176
- Tolerance and Acceptance Within Herbalism — 177
- Teaching Through Experience versus Lecture — 179
- Classroom Learning and Apprenticeship Experiences — 179
- Disappearance of the One-on-One Apprenticeship — 181
- What the New Herbal Consumer Needs to Know About Herbs — 181
- Integrating Environmental and Ethnobotanical Concepts into Herbal Education — 183

Chapter 14. James Green — 185
- History of American Herbalism — 185
- American Herbal Renaissance — 186
- Use of Science and Folklore in American Herbalism — 187
- Endangered Plants — 187
- Function of "Medicine" in Healing — 188
- Corporatization of American Herbalism — 189
- Monopoly of Science — 190
- Folklore in American Herbalism — 190
- Nature of Holistic Healing — 191
- Power of the Individual in the Healing Process — 192

Therapeutics in the Marketplace	193
Endangered Plants, Monopopularizing, and Bioregionalism	195

Chapter 15. David Hoffmann — 199

History of American Herbalism	199
Essence of American Herbalism	200
Introduction of Science into American Herbalism	203
Changes in American Herbalism	204
The FDA and the Medical Establishment	204
Herb Industry	206
Herbs versus Herbalism and the Herb Industry's Position on This Issue	207
Speaking the Language of Science	208
Expansion and Potential of American Herbalism	209
Place of Science Within the American Herbal Community	210
Practicing Holistic Herbalism	211
Licensure	212
Intellectual Property Rights	214

Chapter 16. Ellen Evert Hopman — 217

Coming to and Learning About Herbalism	217
Teaching Herbalism	219
Values She Communicates to Her Herbal Students	220
Western Scientific Training for Herbalists	221
Role of the Contemporary Herbalist	222
Medical Science and Its Interface with Herbalism	223
Essence of Herbalism and the Necessary Connection to the Green World	224
Licensure	225

Chapter 17. Steven Horne — 227

Path to Herbalism	227
Nature of Herbalism	228
Herbal Self-Care	228
Work As a Clinical Herbalist	229
Licensure	230
AHG Guidelines	232
Science and Folklore in Herbalism	233

Holistic Herbalism	234
Predictability of Herbs	236

Chapter 18. Karta Purkh Singh (K. P.) Khalsa — 237

Journey to Herbalism	237
Herbs versus Herbalism	238
Consequences of a Lack of Herbalism in America	239
Herbalism in Europe	240
American System versus the European System	241
Benefits of Regulation	242
Licensure and Professionalization	243
Personal Reasons for Supporting Licensure	244
Possible Rifts Within Herbalism Due to Licensure	245
Health Consulting in the Natural Food Store	247
Need for Consumer Education	248
Isolation of Herbalism as a Healing Modality	250
Current AHG Vision of Professionalizing	251
Overall View of American Herbalism	253

Chapter 19. Corinne Martin — 255

Definition of an Herbalist	255
Interface of Herbalism with Medical Science	256
Science and the Tradition of Intimacy in Herbalism	257
Education for Herbalists	259
Integrating Ideas of Intimacy and Spirituality into Herbal Education	259
Providing Alternative Health Care to the Mainstream	260
Regulation of Herbs	261
Licensure for Herbalists	261
Community Garden Project	262
Future of Herbalism	264

Chapter 20. Annie McCleary — 267

Intention and Infusing Spirit into Work	267
Holistic versus Allopathic Perspectives	269
Herbs As a Lifestyle	270
Organic Growing of Herbs, Wildcrafting, and the Link Between Spirituality and the Respectful Use of Plants	270
Her Apprentices	272
Ritual and Infusion of Spirit in Daily Life	272

Chapter 21. Michael McGuffin — 273
- Holistic Healing — 273
- Holistic Healing in the Marketplace — 274
- Responsibility of the Consumer — 275
- Herbs versus Herbalism — 276
- Proper Use of Herbal Products — 277
- Endangered Plants — 278
- Monopopularizing — 280
- Folklore in the Market Setting — 281

Chapter 22. David Milbradt — 283
- Definition of Herbalism — 283
- Role of the Herbalist — 283
- Licensure for Herbalists — 284
- Science and Folklore in Herbalism — 286
- Push for Clinical Trials on Herbs — 288
- Allopathic versus Holistic Herbalism — 289

Chapter 23. Orest Pelechaty — 291
- American Health Care and the Place of Herbalism Within It — 291
- Role of the Herbalist — 293
- His Own Experience Developing a Practice of Health Care — 295
- Paradigm Shift in the Mainstream and the Potential of Herbalism — 296
- Licensure and the Potential "Softening" of Science — 297

Chapter 24. Deb Soule — 299
- Observations of Endangered Plants — 299
- Bioregionally Endangered Plants — 300
- Role of Alien Species in the Loss of Habitat — 301
- Role of the Herbal Industry in Endangered Plants — 302
- Gardens at Avena Botanicals — 303
- Our Culture's Relationship with Earth — 303
- Domino Effect with Endangered Plants[4] — 304
- Holistic versus Allopathic Use of Herbs and Education for the General Public — 305
- Variety and Balance in the Herbal World — 306
- Her Wish for Herbalism — 306

Chapter 25. Sharol Tilgner — 309
- Her Work with Wise Woman Herbals (WWH) — 309
- How WWH Began — 310
- The Audience for WWH — 310
- Current Herbal "Boom" — 311
- Sources for Herbs and Endangered Medicinal Plants — 312
- Identifying Plants for Use in Medicines — 313
- Federal Regulations and Small Businesses — 314
- Roles of Science and Folklore — 315
- Bringing Holistic Concepts to the Mainstream — 317

Chapter 26. Peeka Trenkle — 319
- Holistic and Allopathic Herbalism — 319
- Herbal Education — 321
- Spirituality in Herbal Healing — 322
- Herbalism As Consumerism — 323
- How Do We Get to Her Ideal Herbal Vision? — 324
- Licensure and Legitimization — 325
- Potential for Rifts in the American Herbal Community — 326
- Moving into the Mainstream — 327

Epilogue — 329

Appendix A. United Plant Savers "At-Risk" and "To-Watch" Plants — 333

Appendix B. Resources for Further Study — 335

Notes — 339

Bibliography — 345

Index — 353

Foreword

This book is the first of its kind—a boundary breaker between allopathic and herbal medicine. Anne Dougherty clearly illustrates how herbs can be used in the same way as allopathic drugs—this herb for that condition—but goes far beyond that "botanical" paradigm to describe the myriad possibilities for integrating herbalism into everyday life in contemporary technocratic society. She expresses acute awareness of the dangers massive industrial exploitation poses to the plants themselves, and of the importance of realizing that the "whole plant," when picked and utilized with conscious appreciation and awareness, can have healing benefits not realized under the industrial paradigm. Herbs, like people, cannot be reduced to their constituent elements without the loss of the power of the whole. Dougherty makes this clear. For her, the "whole" means the whole plant, the whole person, and the whole of nature. Together, these can produce a stunning synergy of healing; separated, herbs can become no more than replacements for allopathic drugs.

I suggest that this difference between the whole and the part is best understood in the context of the three primary paradigms of health care in the contemporary Western world—the technocratic, humanistic, and holistic models.

The Technocratic Model of Medicine. As with all health care systems, the medical systems of the industrialized world embody the biases and beliefs of the societies that created them. The word I use to describe such societies is "technocracy." A technocracy is a society organized around an ideology of technological progress. What most people call "the medical model" I call "the technocratic model of medicine," or "the technomedical model," to make clear the connections between this model and the core values of the industrialized societies that created it. The main value underlying the technocratic paradigm of medicine is separation. The *principle of separation* states that things are better understood outside of their context, divorced from related objects or persons. Technomedicine continually separates the individual into component parts, the process of medical treatment into constituent elements, and the herb into its "essential element," meaning the extraction of the supposed one active element from the whole plant.

In this way, the technocratic model separates the human body from the human mind and defines the body as a machine—a metaphor that reflects the technocracy's emphasis on the importance of machines and on the resulting value of extracting the appropriate "oil" or "lubrication" to keep the machine running properly. Mechanizing the human body and defining the body-machine as the proper object of medical treatment frees technomedical practitioners from any sense of responsibility for the patient's mind or spirit. Concomitantly, this ideology denies any sense of the value of the whole—the whole person, the whole herb. Thus the energy extant in the whole herb becomes irrelevant—it is only the active ingredient that matters.

In technomedicine, diagnosis and treatment are approached from the outside in. When most machines break down, they do not repair themselves from the inside; they must be repaired from the outside, by someone else. Thus in technomedicine, it follows that one must attempt to diagnose problems, cure disease, and repair dysfunction from the outside. The most valued information is that which comes from the many high-tech diagnostic machines now considered essential to good health care.

Technocratic medical systems also rely upon hierarchical organization and standardization of care; in this way these systems constitute a microcosm of technocratic society. They routinely subordinate nonmedical practitioners to MDs, and individual needs to standardized institutional practices and routines. Such standardization can result in the prescription of X drug (plant- or synthetically based) for X condition, often with minimal consideration for the needs of the individual in relation to the properties of the specific drug.

In line with its hierarchical structure, the technocratic model invests authority in physicians and in institutions and their personnel. When the doctor is the authority, the patient abdicates responsibility. And when technology is essential to diagnosis (the CAT scanner, the fetal monitor, the dialysis machine), all other knowledge systems, such as those of herbalists, are discounted.

Furthermore, technomedical systems utilize aggressive intervention with emphasis on short-term results, and consider death as defeat. Since the dawn of the Industrial Revolution, Western society has sought to dominate and control nature. The more we controlled nature, including our bodies, the more we feared the aspects of nature we could not control, leading to what anthropologists have labeled the "one-two punch" of technological intervention. The one-two punch is a powerful motivating force in technocratic societies—I call it *the technocratic imperative*. This impetus to improve on nature through technology has as its ultimate aim to free us altogether from the limitations of nature. The more able we become to control nature, including our natural bodies, however, the more fearful we become of the aspects of nature we cannot control and the less trust we have for remedies that work in harmony with our bodies in lieu of controlling them. Death becomes the ultimate signifier of defeat and thus the ultimate enemy.

Technocratic imperatives combined with a profit-driven system and an intolerance of other modalities have raised the technomedical paradigm to a hegemonic ideal. Hegemony refers to an ideology espoused by the dominant group in a given society. When an ideology is hegemonic, all other competing ideologies become "alternative" to it. Thus, in the United States, healing modalities such as midwifery, chiropractic, homeopathy, naturopathy, acupuncture, herbalism, and so forth have been viewed as alternative to allopathy, which still sets the standards for care. Its hegemonic status works to ensure its profitability: pharmaceutical and medical technology companies constitute by far one of the most profitable industries in technocratic societies. Any system that gains sociocultural ascendancy and then rigidifies, shutting out new information and refusing to incorporate contradictory evidence, is in mortal danger both to itself and to the public it serves. Yet the power of the technomedical paradigm is such that physicians will rapidly accept procedures and technologies in keeping with it, while rejecting those that are not.

The Humanistic Model of Medicine. In the United States and elsewhere, the excesses of technomedicine have long been the subject of heated discussion and debate. Humanism arose in reaction to these excesses as an effort driven by nurses and physicians working within the medical system to reform it from the inside. Humanists wish simply to humanize technomedicine—that is, to make it relational, partnership-oriented, individually responsive, and compassionate. This caring, commonsensical approach is garnering wide international appreciation and support. Clearly less radical than holism, clearly more loving than technomedicine, the humanistic paradigm has the most potential to reform the technocratic system from the inside.

The basic principle underlying the humanistic approach is connection: the connection of the patient to the multiple aspects of self, family, society, and health care practitioners.

The humanistic approach recognizes the influence of the mind on the body and advocates forms of healing that address both. Although in some ways the human body is *like* a machine, it is a fact of biological life that the body is not a machine but an organism. Such a conclusion has powerful repercussions for treatment, as the way the body is defined will shape the way it is treated by a culture's health care system. Defining the body as an organism charters the development of an array of treatments which may be irrelevant to a machine but matter a great deal to an organism. Unlike machines, mammalian organisms feel pain and respond emotionally to interactions with others and to changes in their environment. Most mammals respond positively to the comfort of a loving touch, an understanding conversation, or a nutritional strategy that strengthens their organic ability to resist or heal disease. Thus a paradigm of healing based on a definition of the human body as an organism logically stresses the importance of caring, of touch, and of appropriate nutrition. The best analog for the term *humanism* in the medical literature is the term *biopsychosocial*, which acknowledges that this model takes into account biology, psychology, and the social environment.

Humanism requires treating the patient as any human being would want to be treated—with consideration, kindness, and respect. Humanists work to establish strong connections with their patients and endeavor to know them as individuals. In contrast to technomedical professionals, humanistic practitioners value the input of the patient more prominently using this information to create a balanced health care plan. Humanistic practitioners elicit information from deep within the patient and combine it with objective findings. They find that *how to listen* is as important as knowing what to say, as a patient's story can provide important keys to treatment. And humanism counterbalances technomedicine with a softer approach, which can be anything from a superficial overlay to profoundly alternative methods.

In the technocratic model, the discussion of options outside of conventional medicine is generally impossible due to the doctor's allegiance to technocratic approaches and ignorance of alternatives. But in humanism, open discussion of treatment choices leads naturally to an exploration and sharing of values. Humanistic physicians take science as their standard and use virtually the same tools and techniques as technomedical doctors. The difference lies in timing and selection. Humanists are more willing to wait, to be conservative, more open to mind/body approaches. They use technology, but emphasize caring and relationship alongside it, a combination captured in the phrase "high tech, high touch."

Most proponents of humanism are also strong proponents of disease prevention and science-based public health initiatives. They point out that a clean water supply will do more good for the health of many more people than building high-tech hospitals, as will ensuring clean air, adequate nutrition, and access to primary health care. Thus the public health paradigm, which stresses long-term, large- scale disease prevention, public education, and health promotion, corresponds closely to the humanistic paradigm, which stresses long-term individual and family disease prevention and health promotion. Humanistic medical practitioners often leave private medical practice for work in the wider arena of public health. The public health paradigm takes a broadscale, population-wide approach to disease prevention and material health, while the humanistic model focuses more specifically on the individual relationships between family, patient, and provider, and their effects.

The driving ethos of the humanist is compassion—the ability to sense and care about the needs of others. Most humanists have no intention of learning alternative healing techniques, although in general they are open-minded and support patients who choose to use alternatives. Physicians and other medical practitioners seeking to practice humanistically need not undergo any noticeable change in beliefs about what causes or cures disease. Simply being nicer, more caring, more willing to touch and communicate repositions them in the humanistic model. Most will not undergo the radical shift in values that permits them to go beyond compassion to employ the healing power of that mysterious thing called energy in overcoming disease. This is the realm of the holistic healer.

The Holistic Model of Medicine. The principles of connection and integration that underlie the holistic paradigm arise from the fluid, multimodal, right-brained thinking that, after centuries of devaluation in the West, is finally beginning to regain lost ground. It is thinking of, with, and through the body and the spirit—holistic, fluid thinking that transcends logical reasoning and rigid classifications and encompasses the unpredictable relationship, the unexpected connection, the revealing intuition, the synergy of remedy and spirit—that so often constitutes a prime element of holistic healing.

A large part of the initial impetus for seeing mind and body as one in holistic healing was the realization that the brain, the physical seat of the mind, is not located only in the head but in fact extends throughout the central nervous system. Understanding this makes it much harder to think of body and mind as separate entities. If the mind is the body, and the body is the mind, then addressing the psychological states and emotions of any patient becomes *the* essential aspect of care. The holistic paradigm also insists on the participation of the *spirit* in the human whole. In incorporating spirit or soul into the healing process, holistic healers bring medicine back into the world of the spiritual and the metaphysical from which it was separated during the Industrial Revolution.

The holistic paradigm moves far beyond the narrow view of the body-as-machine, past the humanistic view of the body as an organism, all the way to a limitless view of the body as energy. Defining the body as an energy system provides a powerful charter for the development and use of forms of medicine and treatment that work energetically such as acupuncture, homeopathy, intuitive diagnosis, Reiki, hands-on healing, magnetic field therapy, and therapeutic touch. "Energy medicine" acknowledges that an individual's health can be influenced by such subtleties as the vibrations of anger or hostility. The important point is that for the practitioner who works at the level of energy, these sorts of interventions will not be afterthoughts or overlays, but will be basic and primary—the first line of care.

As we have just seen, the holistic model offers the possibility that the client and practitioner are not separate but are fundamentally one. If the body is an energy field, then as they interact the energy fields of client and practitioner can merge. For example, one midwife entered a labor room to find a young woman rocking on the bed and whimpering "Oh God, Oh God" in a high-pitched voice. The midwife simply climbed onto the bed and held the laboring woman in her arms, rocking with her and whispering in her ear in deep tones, "Oh good, oh good," until the woman began to chant with her and soon her sounds deepened to "Oh, gooo-oo—ood" because suddenly, as she released her fear, her body opened and she was ready to push.

Holism acknowledges that no single explanation of a diagnosis, no single drug or therapeutic approach, will sufficiently address an individual's health problems; rather, such problems must be addressed in terms of the whole persons and the whole environments in which they live. It is no accident that the most commonly asked question in holistic health is "What's going on in your life?" This question expresses the holistic view that illness is a manifestation of imbalance in the body-mind-spirit whole. Holists note that the health of the immune system, or the process of pregnancy and birth, can be impeded by exhaustion, depression, and emotional stress. And they believe that a healthy immune system, as well as a healthy pregnancy and birth, can be facilitated by multiple means, from dialogue to dream analysis to dance, from massage to exercise to whole herbs and organic food.

While they may, if appropriate, order "outside-in" diagnostic tests, holistic practitioners will primarily diagnose and treat from the inside out—in other words, they will rely to a significant extent on their own intuition and the inner knowing of their clients as primary sources of authoritative knowledge, along with the books and the machines. Their willingness to rely on intuition comes from their deep understanding of the body as energy and their trust in right-brained, gestaltic kinds of thinking that do not rely on logic but on that sudden flash of insight from which healing can arise.

A basic tenet of holistic healing is that ultimately, individuals must take responsibility for their own health and well-being. Holistic practitioners tend to see themselves as part of a healing team, of which the patient is the most significant member. Holistic healers in general do not reject technology; rather, they place it at the service of their clients. Usually their technologies are not invasive, nor do they produce the toxic effects of many of the technologies of technocratic medicine. Such technologies do not dominate and control; rather, they work with physiology to empower the healing process.

Holistic practitioners focus on long-term creation and maintenance of health and well-being and view death as a step in a process. Technocratic practitioners often express extreme frustration regarding the patient's failure to follow doctor's orders. In contrast, holistic practitioners most frequently voice frustration with patients who make no long-term commitment to improving their health but want the healer to provide them with a quick fix. Holistic practitioners want their clients to make long-term changes in their diets and lifestyles that will not simply prevent illness but will actively generate good health. Giving up sugar, caffeine, and highly processed foods, taking herbal and vitamin supplements, eating nutrient-rich organic vegetables and complex carbohydrates, exercising regularly, and dealing with stress through meditation are examples of the kinds of long-term changes that are often necessary to the creation of wellness. Of course, many people are resistant to such lifestyle alterations. Holistic practitioners must engage in a great deal of client education, and must maintain a great deal of patience, in order to support people in making this kind of change.

The holistic paradigm redefines death not as any kind of final end but as an essential step in the process of living. This view stems from holists' definition of the body as an energy field, and from their deep-seated understanding of the transmutable nature of energy. Because of their integrated views on the essential oneness of body, mind, and spirit, it is only at the moment of death that holists grant these a conceptual separation. At death, in this view, the energy of the body decays and returns to earth, while the energy of the spirit or the individual consciousness continues on. Many holists accept some version of Eastern philosophies of reincarnation that see death as an opportunity for continued growth into a new kind of life in spirit and then perhaps again in flesh. While this positive view does not lead holists to rush to embrace death, it does tend to give them a strong sense of trust in the essential safety of the universe and in the wisdom and worth of its ways.

Although holisitic practitioners are conscious of the need to earn a living, it *follows* their personal commitment to work rather than drives it. Recognizing that healing occurs not in response to their actions but in the support and stimulation of the vital force, in the exchange of energy between individuals, or in the long slow progress toward health that often rewards serious lifestyle changes, holistic practitioners are keenly aware of their partnership with patients. Money is part of this exchange. Unlike technomedical practitioners who are apt to live stressful and harried lives wherein they are unable to care for themselves adequately, holistic practitioners find that their own healing often accompanies that of their patients, as it is impossible to fully espouse a holistic philosophy without applying it to oneself. In the mutual appreciation that often arises between holistic healer and patient, a deep experience of *value* replaces the focus on money.

The holistic paradigm's definition of the body as an energy field in constant interaction with other energy fields makes possible its embrace of multiple modalities that remain unacceptable to proponents of technomedicine. The holistic model encompasses a rich variety of approaches, ranging from nutritional therapy to herbalism, to traditional healing modalities such as Chinese and Japanese medicine, to various methods of directly affecting personal energy. Some holistic practitioners study a particular modality while others employ an eclectic approach. The ultimate holistic vision entails a profound revolution in health care that would replace the dominance of the technomedical model with a multiplicity of approaches. Herbalism, midwifery, homeopathy, naturopathy, acupuncture, and other modalities would take their places as respected and legitimate disciplines. Practitioners of each modality would know enough about the others for appropriate referral. Above all, the public would be educated in the techniques of self-care, healthy lifestyle, and the appropriate use of a variety of approaches to healing.

As a society's medical system mirrors its core values in microcosm, so the evolution of medicine can influence the evolution of the wider culture. We must ask, Who do we want to make ourselves become through the kinds of health care we create? Contemporary practitioners have a unique opportunity to weave many elements together. Information is available about indigenous practices from many cultures, some of which are highly beneficial and should be incorporated. More information than ever is available from scientific studies that inform us about the physiology of many conditions and the kinds of care that truly support people to heal.

Herbalism is relevant in all three paradigms. Hundreds of allopathic drugs have been developed utilizing the "active principle" in multiple herbs, with the limitation, which Dougherty stresses, that the healing effects of the whole herb are ignored. Humanistic practice, while still based in a technocratic approach, moves toward a deeper understanding of the body as an organism and thus of the value or organic remedies such as herbs and improved nutrition. And holistic practice requires a full understanding of the value of multiple kinds of remedies, including herbs and their many-faceted values in healing.

Dougherty's own understandings and analysis constitute the first sections of this outstanding and user-friendly book. Illuminating interviews with practicing herbalists constitute the rest. In their words, we read the struggles of those who appreciate nature and seek to bring its benefits to the world with the misunderstandings they encounter, the damage their own work can do to the herbs and the environments within which these herbs flourish, and their efforts to salvage both the herbs and (mis)understandings of the people who use them.

Dougherty's appropriate use of systems theory for explicating individual and cultural relationships to the healing power of herbs extends her analysis beyond her subject, linking it to other such studies of alternative healing systems. Both as an anthropologist and as a frequent user of herbs for my own healing, I am captivated by this book, from beginning to end. I find it to be a work of information, holistic comprehension, and deep communication between the worlds of herbs, herbalists, and the wider society that is increasingly coming to appreciate and rely on their healing powers.

Robbie Davis-Floyd, PhD
Adjunct Associate Professor
Case Western Reserve University
Cleveland, Ohio
and Senior Research Fellow
Department of Anthropology
University of Texas, Austin

Preface and Acknowledgments

American herbalism is a field that cherishes and honors longstanding tradition and those elders who have built that legacy. Respect for elders is one of the first axioms of American herbalism that I realized. With that notion in my head, it was initially difficult to pose challenges to the thinking of my elders in the contemporary community. It became difficult to say, "Wait a second. What is this grounded in and where are we going as a community?" A good friend reminded me, though, that just as today's longtime herbalists built upon the tradition they were handed, they should and do welcome younger herbalists taking an interest in and questioning the present tradition. The common goal is to make American herbalism a sustainable, thriving healing art and science, a relevant part of modern life and health care.

With that understanding, I was freed to approach highly experienced clinical herbalists and herbal educators; I was freed to ask the ponderous and challenging questions about the nature of holism and what it meant to be truly integrative. Throughout the process, the herbalists I interviewed were encouraging. They spoke frankly about their joys, fears, and the places that housed the most concern for them. Without their voices, resonant and clear; their willingness to share; and their ability to accept the critique of a summertime herbalist (as James Green calls the new generation), this project would have remained an ungrounded and spiraling set of questions in my head. Although not all interviews' transcripts are included in this volume, all of the herbalists with whom I spoke informed my perspective. My deepest thanks and gratitude go out to those herbalists, holders of a precious tradition, who responded to my call: Laura Batcha, Howard Buckingham, Richo Cech, Ryan Drum, Gayle Eversole, Daniel Gagnon, Leslie Gardner, Kate Gilday, Rosemary Gladstar, James Green, Scott Harris, David Hoffmann, Ellen Evert Hopman, Steve Horne, Patricia Kyritsi Howell, K. P. Khalsa, Efrem Korngold, Corinne Martin, Annie McCleary, Michael McGuffin, David Milbradt, Orest Pelechaty, Carmen Reyes, Nancy Scarzello, Deb Soule, Tom Stewart, Sharol Tilgner, and Peeka Trenkle.

Once I captured the words of the herbalists on tape, the arduous work of transcribing and compiling began. Throughout this process, many friends, colleagues, and advisors helped me transcribe the interviews, then edit, slice, and dice, and finally synthesize it all into the work that follows. Among them, many thousand thank-yous go out to Suzanne Richman, Joyce Kornbluh, and Eleanor Ott whose encouragement and challenges were invaluable, helping me to give clear, organized form to the swirl of thoughts in my head. Special thanks also goes out to my good friend and colleague Rebecca White who listened patiently to endless tales of interview woes, transcribing woes, and doubts in my head and who continually supported me where I was. I would also like to thank the Masè family for their support and love throughout this process. Without the staff at The Haworth Press and my editor, Ethan Russo, this book would not have been possible. I could not continue this list without referring to my siblings, Michael and Mary Clare, and my parents, Joe and Mary Anne Dougherty. They live in my heart and propel my mind. Special thanks go to my father who initially edited hundreds of pages of rough interview transcripts.

With the deepest depths of my being, I wish to thank my partner, Guido, who has journeyed with me many miles through sun, snow, downpours, and droughts. I thank him for continually helping me lift the fog and realize that the sun shines on us if only we will acknowledge it. Ti amo.

Finally, I would like to thank the spirits of the green world who called out to me, grabbed me, and will not let me go. I dedicate myself in service to them and their preservation.

With the blessings of the green, read on!

PART I:
THE WEAVING

Chapter 1

The Dream: My Process and Methodology

It was foggy—so foggy I could not see my hand stretched out at arm's length in front of me. I continued walking even though I did not know where I was or where I was going. I had only a vague sketch of who I was, only scattered pieces of my desires, my happiness. It was almost as if a five-year-old child was trying to complete a thousand-piece puzzle. She did not grasp the full picture and so struggled along with the tiny slivers of the project.

The dream was not scary. It felt oddly comforting to be without sight and know only that X and Z made me feel happy and fulfilled while Y did not. I could become an observer of my life pattern, letting desires and adventure pull me here or there. If it was meant to be, it would happen. If not, well . . . off I go.

At a certain point, my partner, Guido, appeared alongside me. We both walked without sight, but so closely to each other that we could see the other's face clearly. We walked in a strong embrace, intuitively understanding that we were heading in the same direction although we could not assign a geography to it.

I could feel soft dewy grass beneath my feet; I could feel the velvet moss pad my way as I walked. I recognized at an early point that concrete and pavement were not part of my reverie. Although I was not sure where in geographic space I was, I was certain it was raw wilderness. I have always felt a strong pull toward Earth. It is part of my nature. I am a Capricorn. The Earth wisdom flows through me. It is in the stars. It is my destiny.

As I wandered farther, the ground became more visible. I began to discern green plants under my feet, brushing against my bare ankles. At first, they were simply a sea of lush green, indistinguishable from one another. As I walked on, however, I began to recognize their unique forms. I understood that they grew in family or community units. The spirits of the plants called to me; their essences addressed me by name.

"Anne, recognize me. I am Lord Yarrow, overseer of the expansive meadow. I shine blindingly in the pure sunlight. You will see that later."

"I am Saint John. Notice the pinprick holes in my leaves. I let the light come through. Admire the warmth of my yellow flowers. Look for me in the fields and allow me to propagate."

"Anne, see me now. I am soft Mullein. Use me as a pillow. I soothe. You will no doubt recognize me from afar. I stand tall and proud."

"Look, but do not touch me, Anne. I am fair Lady Slipper, elusive and rare. Observe my sensuality. Learn what you can."

I heard their voices float through the air. I heard their voices and did not understand why no one else did. I walked miles into the fields and forests of my new plant friends; I met many more as I journeyed through the fog. As my connection deepened, I knew I had to be involved with them each moment of each day. With every step, I understood more how my connection to Earth could manifest in caring for these plants, striving to understand their essences and characteristics, but I could not comprehend why, as I felt closer to my truth than ever, the fog was still so thick.

One day, at the height of my frustration, I thought I would try to strike a deal with the plants and employ their power. I consulted with representatives of the plant community and said, "I will dedicate myself to protecting and caring for you if you will help me to lift this fog."

They laughed as if they were in the know and I, clearly, was not. The chorus answered my plea saying, "Close your eyes as tightly as you can and then open them again."

I did so and was dumbstruck to see that much of the fog was gone. A field of wild herbs shone in front of me. I could see that the council of representatives sat in the center of a large circle of green plants all facing inward. In the distance, I discerned a small red farmhouse.

"You called this into being. This is your land. That is your house. You are the caretaker and steward of this place. Without actively pursuing the idea, you made it happen. You trusted yourself and felt comfortable without your sight. For that reason you manifested this place. This is your unconscious power. Now that you are here, you must continue to lift the fog that is left in the far distance. You must develop your conscious power and actively call things into being. There is much work for you ahead."

I turned and walked toward the small house where Guido was sitting, perched on a rock. *What is conscious power?* I thought to myself.

What I have gathered since the day of my meeting with the plant spirits, the day the fog lifted, is that conscious power is an active form of energy centering. The thought occurs: *I would really like to study herbal healing with a knowledgeable teacher.* I act on that thought. A female co-worker teaches herb classes. Then the reservation comes into play. Those classes are usually quite expensive. I ask the herbalist friend nonetheless. My conscious power overcomes the reservation and therefore presents a realistic possibility. "I am writing a new textbook for my class," she responds, "and if you would help edit it, you could take the class at a sharp discount."

Conscious power is a balanced combination between action and focused attention. The action is essential. To consciously manifest someting I must expend energy in a concentrated direction. As I focus my energy I raise my vibration to the necessary point where "chance" or "fortune" take over. Sometimes it is as simple as making the necessary contact. It may go as far as constructing the appropriate structure or setting. Conscious power requires a willfullness and a direction. One must become involved in order to utilize this power.

Unconscious power, on the other hand, requires an uninvolved yet sensitized process. In other words, one observes an idea or force at work, is aware of it, but does not become entangled with it. To sensitize oneself, one must tear the defensive, protective walls down and function simply on trust and intuition. It is not passive so much as it is uninvolved. I think of it sometimes in terms of a beginner meditation exercise that I was taught many years ago. In the exercise, I was told to relax and envision a river flowing in front of me. Boats would float by that represented thoughts of the mind. I was to observe the boats, or thoughts, but not attach or involve myself with them. I watched them go by but always stayed focused on the section of the river directly in front of me. I was to remain open to the thoughts but not talk back to them. Utilizing unconscious power follows much of the same process. It does not require energy expenditure or willfullness, as does conscious power, but rather surrender. One must allow thoughts, ideas, forces to rush over and through one's being. Through relaxation and trustful surrender, the door to another realm of possibility opens, one not of the rational mind, but far beyond.

My role as a heart-centered herbalist and Earth steward came out of my unconscious power to call things or situations to me. The pure fact that I remained comfortable and open in my journey through the fog set the appropriate movements into action. Each moment of each day, I became more sensitive to what this intuition told me, where it guided me. I listened to the rhythm of Earth and understood more how to harmonize with it.

When I talk about how I got into herbalism, I often explain that I just fell into it. That is true, but clearly I walked myself to the top edge of the garden wall . . . and then fell into the plant world. I took myself (through my conscious power) to the place where the wind needed only to give me a gentle shove.

Looking back on my experiences and ahead on my life path, I am convinced of the necessity of both conscious and unconscious power, both action and meditation. The American herbal community, which I discovered has a keen sense of meditation and a strong belief in unconscious power, is only beginning to understand more of action and conscious power. Although individual herbalists may have a sense of both and a balance between them, the community as a whole tends toward quiet meditation and passivity in the name of a paradigm different from the mainstream, which generally holds consumerism and Cartesian scientific inquiry above all else. If American herbalism is to become a sustainable and relevant piece of an emerging holistic culture, herbalists, as a community, must strive to balance conscious and unconscious power in a series of critical issues, which potentially compromise the long-term sustainability of herbalism in America.

The American herbal community, an eclectic group with specialized and intimate knowledge of how to utilize plants for healing, is a paradox, however; it cannot be described so easily. For all of its passivity, it is reactionary in nature. It is a reaction to technology and its misuse; it is a reaction to market dependency. Again, although individual herbalists hold nonreactionary, integrative philosophies, the community as a whole has not coalesced an integrative vision. Herbalism is left of center; herbalism is fringe. How does a community that defines itself as outside of the mainstream and in opposition to the teachings of the mainstream come into the mainstream? Can there be no reconciliation? How can one integrate with that which one defines as one's opposition? Is it an impossible, nihilistic question? As the American herbal renaissance grows, these are questions faced by the doorstep of the American herbal community.

I began this project initially out of a desire to be in service to the plant community, exclusively and in isolation. The realization that medicinal plants were becoming threatened due to a variety of herbal and business practices was first on my mind (Gladstar and Hirsch, 2000). I felt called to do something concrete about this issue, to be a part of the solution. I planted goldenseal, black cohosh, and American ginseng on my own land. I made it a point to familiarize myself with those plants that were at risk for being depleted in my area. The concept of bioregional herbalism (the use of local plants in healing local people) became central to my herbal practice. I decided to write about the experiences I had and use them to educate a wider audience of herbalists and herbal enthusiasts.

As I explored how to best educate others about endangered plants, I realized how entangled and complicated the situation was within the herbal community and indeed the wider world. "Curing" the endangered plant situation went far beyond replanting the woods and using products containing organically grown herbs. A large component had to do with thought patterns or paradigms. The American public spends billions of dollars each year on herbal medicines (Brevoort, 1998; Robin, 2002). Plants have become endangered plants in part because of an overzealous public buying and using herbs without a strong concept of Earth connection. The situation has worsened due to a media and industry based in capitalist business practices that encourages mass targeted consumerism. Does the American herbal community conteract these forces with a cohesive holistic presentation of its value system? Or is the community being swept up in the torrent of popularity and mainstream recognition? Herbalism has an understanding of a new way to do things, but as a product of the American culture, it may unavoidably fall back into the old, mainstream paradigm of consumerism.

In simply examining the issue of endangered plants, I began to see how strong the pull was to "think *inside* the box," to remain inside the dominant, mainstream paradigm. After existing on the fringe for almost three-quarters of a century, the American herbal community is thrilled to find acceptance and respect from the rest of society. In response, I widened my view and saw that the issue of dueling paradigms, dominant and fringe, was much larger. It covered almost all of the terrain of herbalism and most especially the area where it directly interfaces with the mainstream. A notion is found within herbalism that in order to interface with the mainstream, one must speak and act as if one is part of it, that one must become the mainstream. In doing so, herbalists risk denying or underplaying the beauty and uniqueness of their herbal practices.

Two years ago, as all of these thoughts ran through my head regarding endangered plants and showing how connected this issue is to many others, I sought a methodology for my exploration. After reading Penfield Chester's engaging oral history on American midwifery, *Sisters on a Journey* (1998), I decided that interviewing herbalists would be the perfect way to approach my growing number of questions and observations about the American herbal movement. I would go to the herbalists and ask them what they thought about herbalism. As herbalism has largely been an oral tradition, it seemed only appropriate that I use an oral, storytelling methodology. Furthermore, not much has been written on critical issues in contemporary American herbalism. A few activist-oriented herbalists have written a segment in their books on the subject, but the diversity of viewpoints is not evident in the written material. Therefore, I decided to speak with a diverse array of herbalists. I wanted many points of view within the American herbal community.

My interviews began with prominent teachers who had influenced my development as an herbalist. I asked them about the history of the movement, its fringe nature, and its potential. As I conducted more interviews, the larger scope of critical issues crystallized for me and I redirected my questioning. My inquiries focused on these topics with a short series of questions pertaining to each. The questions remained open-ended, but organization by topic gave me a way to direct the process. The following text considers four of the topics on which I questioned herbalists: science and folklore, licensure, endangered plants, and the place of spirituality in herbalism. Interview topics also included discussion of teaching styles, the preservation of small herbal businesses, education in the marketplace, and holistic versus allopathic uses of herbs. Some of the dialogue on the latter topics has woven itself into the discussions that follow, but due to space constraints, this text does not fully contextualize those issues.

After six or seven interviews with nationally known teachers, I decided to open up my search to full-time herbal clinicians, gardeners, and business owners. I requested interviews with professional members of the American Herbalists Guild, a national organization working toward the professionalization of clinical herbalists. I offered several topics that we could discuss and only asked questions that applied to the preselected topics. Many herbalists responded enthusiastically, and I spent the remainder of my interview period talking with those clinicians and educators.

In the end, I spoke with twenty-eight herbalists from across the country, a diverse group of individuals whose commonality was that they identified themselves as American grassroots herbalists. They lived and practiced throughout the country from Vermont to New Mexico to Oregon to New Jersey to Georgia. The interview participants represent a broad cross section of the non-native American herbal community. Some were avid gardeners and others preferred to walk in the woods. Many of the participants were full-time clinicians, but some were herbal products manufacturers, authors, or educators.

Twenty of these interviews were edited into a monologue format and served as the primary source of information for the discussion that follows. In editing, I omitted my questions so that the narrators' words flow as if they are talking to the reader. The monologues are organized by topic denoted by subheadings. In most cases, the order of the subheadings directly corresponds to the order in which topics were discussed during the interview. In a few conversations, however, an herbalist returned to a topic discussed earlier in the interview with further comment. In such instances, I took the liberty of taking that section of text out of sequence in order to group it with other comments on a similar topic. I have included the edited interviews in the latter half of this volume in the hopes that others may seek out relevant information from them and expand and enrich the dialogue.

This work is a beginning. The interview process not only provided insight into the mind, heart, and practice of the American herbalist but also sparked deeper discussion of many of the issues within the herbal community. There is much more work to be done, however.

With the protective white light of the traveler surrounding us, let us begin our journey into the world of the American herbalist.

Chapter 2

The Terrain: Consideration of Botanical Medicine and Holistic Herbalism

The beautiful thing about herbal medicine is that it has so much aesthetics to it. It has so much life force to it. At its center, herbalism is so much about getting in touch with beauty and using all of your senses—touching and smelling and tasting and feeling. If you go to an herb store, pull a bottle of herbs off the shelf, and open it—and look at it and smell it and taste it—you've gotten so much more involved with that medicine than with a tincture which you just put in drops of water. The more you involve your senses in the process, the more you are enlivening yourself. (James Green, interview, 1999)

James Green was formerly the director of the California School of Herbal Studies, one of the largest herb schools in the country. His voice is clear and poignant as he speaks about reconnecting with nature, returning our beings to balance and harmony, and allowing ourselves to experience the beauty of herbs. If you want to get at the core of herbalism, speak with James. You will float home on a cloud of chamomile blossoms.

As we enter the twenty-first century, the desire of the American public to go beyond the dominant allopathic care system has been realized. Numerous reasons are cited from baby boomers "wanting to seize control of their medical destinies" (Greenwald, 1998, p. 61), to people simply being frustrated with the "cost and consequence" of mainstream health care (Brody, 1999, p. D1). Regardless of the reasons, it is estimated that one-third of all Americans have tried an herbal product of some sort, and the herbal product industry grosses an estimated $3.3 to $4.0 billion per year (Johnston, 1997; Blumenthal, 2002; Brevoort, 1998). According to a Harvard University study (Eisenberg et al., 1998), use of alternative treatments by American adults increased from 33.8 percent in 1990 to 42.1 percent in 1997. Although dips in sales may occur, it is clear that alternative therapies including herbs have entered the mainstream.

With interest in herbal medicine at a 100-year high and money flowing from the pocketbooks of average Americans who are purchasing an array of herbal healing agents, several questions arise. First and most compelling, what is the connection between the billions of out-of-pocket dollars spent and James Green's full sensory description of herbalism? Then, what are and what can be the effects of the current herbal renaissance on our Earth, on the current biomedical paradigm, and on our relationship with and experience of nature?

To begin a discussion about American herbalism and the current herbal renaissance, we must first understand what each of these terms connotes. A delineation needs to be made between two approaches: (1) *botanical medicine*—the use of herbal products as an alternative to mainstream, pharmaceutical medicine and (2) *holistic herbalism*—the pursuit of herbs within a larger lifestyle-altering context. To look at this more deeply, let us first briefly examine an already existing model of healing practices in America.

Through interviews with forty physicians, most of whom consider themselves to be holistic, Robbie Davis-Floyd and her co-author Gloria St. John trace the transformative journey of these doctors in *From Doctor to Healer* (1998). The work delineates a continuum of medical paradigms in contemporary America and attempts to characterize three main points on the line for deep study: the technocratic, humanistic, and holistic (Figure 2.1). The interviews and their analyses not only clearly characterize each phase or point on the continuum but also suggest the possibility of movement from the technocratic to the holistic end while absorbing and adopting the useful tools of stages

Technocratic   Humanistic   Holistic

FIGURE 2.1. Medical Paradigm Continuum (*Source:* From Davis-Floyd, Robbie and Gloria St. John, *From Doctor to Healer: The Transformative Journey.* Copyright © 1998 by Robbie Davis-Floyd and Gloria St. John. Reprinted by permission of Rutgers University Press.)

along the way. Each point on the line builds on the one before, thereby enriching the practice. For example, the technocratic paradigm views the body as a machine as per Cartesian thinking. The job of the doctor then is that of a mechanic who replaces broken parts and lubricates the system.

On the other hand, the holistic paradigm refers to the human body as an "energy system interlinked with other energy systems" (Davis-Floyd and St. John, 1998, p. 117). Although the technocratic physicians have held the view that the body is a series of parts with certain relationships internal to the organism, holistic physicians extend the influence of relationships to include seemingly external forces. A German biologist, Ludwig von Bertalanffy (cited in Capra, 1996), suggested that living entities be called "open systems" to highlight the flow of materials between the organism and its environment. In this view, the organism (the system) and its environment (the surroundings) are not separate but rather conjoined in a much larger system in which reciprocal relationships occur. In this description, the body is fluctuating, pulsing, and alive, in constant relationship with its physical, social, spiritual environments. It is also in relationship with other systems or beings. Since the definition of the system is much larger, there are more points at which disease can arrive and, on the flip side, where health can be affected. The more avenues for affecting health, the deeper the healing may be.

Davis-Floyd and St. John (1998) consider physicians in their study; this study considers American herbalists and their community. Initial consideration of the Davis-Floyd and St. John model brought me to the conclusion that just as there are different types of physicians, there are different types of herbalists. Although numerous approaches to herbal healing are seen in America, including Wise Woman, bioregional, vitalistic, Eclectic—the list could go on ad infinitum—what broad strokes unite and divide these schools of thought and practice? Each falls under the umbrella of American herbalism, but each practices according to a slightly different ethic and ideal. American herbalism is so diverse that it is not easily categorized, and much crossover occurs between approaches, which lends a blended quality to the interior landscape of the field. For the purposes of this discussion, I wish only to discuss in detail two broad subsets within American herbalism—botanical medicine and holistic herbalism. With an understanding of these practices, discussion of critical issues faced by the American herbal community may be more easily contextualized.

As I thought more about the qualities of each thread or paradigm within American herbalism, I realized that they overlap the Davis-Floyd and St. John (1998) categories in some ways. Both botanical medicine and holistic herbalism utilize pieces of the technocratic, humanistic, and holistic paradigms as presented in the established model. For that reason, I propose an overlay of the Davis-Floyd and St. John (1998) model, with botanical medicine straddling the technocratic and humanistic paradigms and wholistic herbalism spanning the holistic and the space beyond it. Figure 2.2 provides a schematic of this concept. A listing of the characteristics of both botanical medicine and holistic herbalism as well as an in-depth discussion of each point follow.

BOTANICAL MEDICINE

1. Specific herbs for specific healing
2. Exclusive use of scientific exploration in evaluating herbs
3. Mind-body connection
4. Earth connection in dormant seed phase

Botanical medicine, although relatively new to the field of American herbalism, is presented most prominently to mainstream America through the herbal renaissance. It utilizes language and approaches similar to the biomedical model, a healing paradigm familiar to modern Americans. In that way, botanical medicine is understandable and graspable to large numbers of Americans who seek alternatives to the

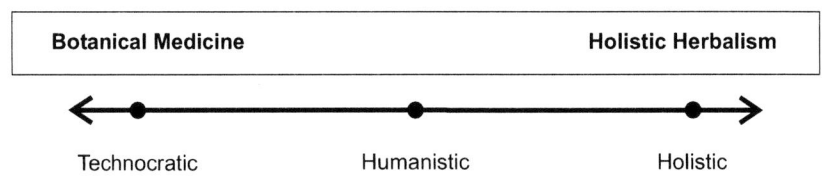

FIGURE 2.2. Overlay of Davis-Floyd and St. John (1998) Model and Medical Paradigm Continuum (*Source:* From Davis-Floyd, Robbie and Gloria St. John, *From Doctor to Healer: The Transformative Journey.* Copyright © 1998 by Robbie Davis-Floyd and Gloria St. John. Reprinted by permission of Rutgers University Press.)

biomedical model but have limited knowledge as to how these alternatives work.

Specific Herbs for Specific Healing

In today's mainstream, the use of botanicals is strictly a "this for that" application, a particular herb for a particular disease or health issue. In a 1998 issue, *Time* magazine (Greenwald, 1998) dedicated six pages to color photos of the hottest botanical medicines. At the bottom of each page the use for each plant was highlighted: St. John's wort for mild to moderate depression, garlic to lower cholesterol. A quick scan of herbal product labels shows that ginkgo is marketed toward those with a fading memory and kava to the anxiety ridden.

Within the biomedical model of health and disease, a patient comes with an ailment, and a doctor prescribes an appropriate pharmaceutical to ameliorate or cure the condition. Particular pharmaceuticals treat particular conditions. As people move from this allopathic paradigm into botanical medicine, much of the approach remains the same, but the choice of ameliorative substance changes from "drug" to herb. For example, within botanical medicine, herbs such as red clover or black cohosh are used to replace hormone therapy for menopausal women. There is a specific herb for a specific condition, or as Ryan Drum (interview, 1999), professional clinical herbalist, says, there are, "standardized patients for standardized medicines." Peeka Trenkle (interview, 1999), another professional herbalist, sums up the situation:

> What is being called holistic is just using herbs the way drugs are being used. It is using the same medical construct and instead of using nystatin for yeast they are using calendula or thuja or monarda. It is still a certain classification of herbs for a certain classification of disease.

As we move across the continuum from the allopathic into the botanical medicine paradigm, we have broadened the available remedies but have remained within the biomedical clinical context: the body is a closed system (the surroundings have no discernable effect on the system), and herbs are natural drugs that we use to affect the healing process or status quo maintenance of the body.

Exclusive Use of Scientific Exploration in Evaluating Herbs

In a 1999 keynote address to the American Society of Pharmacognosy, well-known pharmacognosist Varro Tyler, now deceased, spoke of the need for scientific research on plant medicines. Without such research, it was his heartfelt belief that herbal medicines would fail to play a major role in modern health care. He went on to emphatically state that any who would disagree with the need for scientific validation were out of touch with reality (Tyler in Skinner, 1999, p. 68).

Science is the authoritative language of American culture. If it is scientifically proven, it must be real. If it is real, it must be able to be scientifically proven. Tyler's dramatic words demonstrate the intense demand for scientific inquiry into herbs. David Hoffmann (interview, 1998), well-known clinical herbalist and teacher, explained where some of the drive for scientific studies comes from.

> If the media is going to look at herbs for a feature article or something, it sounds a lot better if they get the information from an ND [doctor of naturopathy] or a pharmacologist. Journalists have no idea what the issues are. They just pick up on the science.

Furthermore, many who work with herbs have a desire for their work to be accepted by a large-scale audience. To do that, they feel that they must use scientific inquiry to the exclusion of all other modes of explanation. Again, David Hoffmann notes (interview, 1998), "The only way people can ever hear anyone else is if the person is expressing the issue in their language." In the case of addressing herbs to the American majority, that language is science.

As will be discussed in more depth later, the study of herbs and their uses holds a wealth of empirical information based on generations of traditional use of plants. Many botanical medicine practitioners and teachers, however, do not avail themselves of the rich historical evidence for the efficacy of herbs. Although an understanding of the biochemistry of phytomedicines enriches herbal practice, the lack thereof does not indicate that the herb has no effect. Media reports make it seem as if no one is quite sure what a particular herb does, therefore rendering it unworthy of consideration. The fact of the matter may be that the documentation of the herb is not in modern scientific language and is, therefore, considered irrelevant. Skullcap is such an example. Due to a lack of clinical trials and pharmacological work on *Scutellaria lateriflora,* practitioners of botanical medicine virtually ignore this herb, which is prized by many practitioners of holistic herbalism as a superior nervine tonic (Low Dog, 1997). In contrast to holistic herbalism, which places high value on traditional evidence, botanical medicine is much more comfortable with the indications derived from modern clinical trails. To follow up the *Scutellaria* example, the pressure for scientific validation of its effects is so intense that a group in California (including herbalist David Hoffmann) has recently finished a placebo-controlled, double-blind study on the herb (Wolfson and Hoffmann, 2003).

Botanical medicine in its attachment to science balances between the biomedical world and that of holistic herbalism. Its exclusive use of modern clinical trials as proof of efficacy allows it to appear well defined and solid in contrast to the more traditionally based, experiential understanding of holistic herbalism. The subject of science will be discussed in greater depth in Chapter 3.

Mind-Body Connection

Each step along the continuum adds a facet to the health/healing complex. Biomedicine begins by considering the body an entity to be "fixed." Botanical medicine goes the next step to include the mind—the mental and emotional health. This paradigm understands that the mind can have effects which manifest in the physical body. In the botanical medicine paradigm, however, the mind is separate but equal. That is, the mind is recognized as an important component of health, but its influence on the body is not fully realized.

Andrew Weil, in his best-selling book, *Spontaneous Healing* (1995), begins to communicate this idea to the mainstream public. He relays stories of guided imagery to directly illuminate the effect that the mind can have on the physical realm. In one story, his wife undergoes a form of mind-body therapy to deal with pain in her lower back. While in a trance state, she relates to the practitioner that her back is angry with her for using cold compresses rather than hot to alleviate the pain. The practitioner tells her to agree to switch to hot therapy and ask if she can make peace with her back. All parties agree to make peace, and after regaining normal consciousness, Weil's wife is free of back problems before she even uses the promised hot compresses.

Although botanical medicine is progressive in its recognition of the role of the mind in healing, it does not necessarily understand the two to be unified. Andrew Weil's example shows one aspect (mind) affecting another aspect (body). This cause-and-effect relationship indicates a separation between the two. As will be discussed under the paradigm of holistic herbalism, it is possible to eliminate that separation and recognize a oneness about the system. It does, however, become quite esoteric and difficult to speak about through rational analytical writing (Starhawk, 1989).

Earth Connection in Dormant Seed Phase

> The Herbal Product Industry makes [several billions of dollars], and what part of that goes through the hands [of the herbal practitioner]? Not much. What part of it goes through retail setting? Almost all of it. What part of that part in the retail setting is touched by a deep knowledge of [Earth-centered] herbalism? Some, not much. So I think that what's being marketed is herbs—herbal products, herbal formulas. Nobody's marketing herbalism [and connection to Earth]. (Michael McGuffin, interview, 1998)

The benchmark of the American herbal renaissance is marketplace dollar figures. An article in *HerbalGram* states that between May 1997 and May 1998, sales of herbal products increased 101 percent (Anonymous, 1998). Although certain sectors have experienced periods of slower growth since that time, the natural food market continues to grow. A recent survey by *Prevention* (cited in Blumenthal, 2002) showed sales of single-herb extracts equaling $1.34 million in 2001 as compared to $1.23 million in 2000. What is being marketed are herbal products, not the aesthetics and spirituality that James Green mentioned at the beginning of this chapter. The connection to Earth as the source of plant medicines is in the dormant seed phase; it is there underlying the trend, but not expressed as such.

Botanical medicine uses plants as healing agents, but the realization that the medicine comes from the Earth is quiet. That is to say, most practitioners of botanical medicine obtain their herbs from a health food market or corner drugstore in the form of a capsule, pill, or standardized extract with little or no contact with the live plant.

Deb Soule (interview, 1999), herbal teacher, writer, and gardener, passionately explains:

> At this point, we are in a time where the bottom line for so many people is money. The relationship to the natural world and to the health of the ecosystem as a whole is not at the top of the list for most. . . . The faddish part of herbalism is dangerous as far as people having no relationship. They don't have a clue as to what these plants look like to begin with and no, or little, relationship to the fact that these are amazing living beings that have as much of a right to be here as we do. [And that realization] helps to keep their environment safe and flourishing in a good way. Most consumers, unless they have the opportunity to take the time to really look deeply and educate themselves, . . . don't really understand it.

Regardless of the phase or stage of the connection to the Earth, most herbalists support universal access to herbs for those who want to use them as part of their health care program. Corinne Martin (interview, 1999) said,

> I think that accessibility for the mainstream is really important. I wouldn't withhold an herb from [people] just because they didn't want to live the way I want to live. That doesn't make any sense. If people can start out thinking about plants because they have used it in a capsule or their physician prescribed it or suggested it and it worked for them, then yippee. . . . If people are interested and they can use something that is less toxic, why not? Somewhere along the line they are going to figure out that it is about a plant.

As we think of the characteristics of botanical medicine and holistic herbalism, we must keep in mind that, just as the human body is an open system, so too is herbal healing. The existence of distinct threads of herbalism adds challenge and conflict that stimulate discussion and introspection within each practice and, ultimately, enrich the terrain of herbalism.

Furthermore, each paradigm, as in Davis-Floyd and St. John's model, builds on another. It is not unlike the cyclical flooding of a river. Each point on the line may be considered a season of flooding that leaves behind its silty deposits to enrich the flood plain soil. The inhabitants of the area in turn cultivate the land, which produces greater and greater yields each year. The resulting soil, built of years of silt, will serve to nourish a nation of people.

HOLISTIC HERBALISM

1. Herbs as a lifestyle
2. Earth connection blossoming
3. Grassroots organization and emphasis on independence
4. Inclusion of both scientific and folkloric ways of knowing
5. Body-mind-spirit connection
6. Treat the person/attention to the individual

As we journey along the continuum of American herbalism, past botanical medicine, we arrive in the territory of holistic herbalism. As mentioned earlier, holistic herbalism is a general term that describes an entire lifestyle rather than simply a healing practice. It contains many different approaches and philosophies of life and healing, including Wise Woman ways, biodynamic and bioregional approaches, and some of the newer hybrid "East meets West" fusions (Foster and Chongxi, 1992), which bring together pieces of Chinese, Ayurvedic, and Eclectic traditions, to name a few. It must be noted here that the majority of the herbalists who narrate this story identify with the tenets of holistic herbalism most closely.

Herbs As a Lifestyle

Although holistic herbalism still uses herbs to effect healing in an individual, the conscious use of plant materials reaches beyond pharmaceutical-like applications of herbal medicine as seen in the botanical medicine paradigm. Herbs, thinking about herbs, harvesting herbs —all direct the life of the herbalist. Ellen Evert Hopman (interview, 1999), a teacher of herbalism as well as paganism, describes,

> The herbalist is a lot closer to the Earth and works on a much more basic level. If you are a long-term herbalist, it is probably because you have a very strong spiritual relationship with the

herbs and the trees. They are like your family or you wouldn't keep doing it. You live in the green world.

Indeed, as with family, the herbalist is surrounded by plant material and lives with the rhythm of the Earth's seasons. Rosemary Gladstar (interview, 1999) suggests that those interested in herbal medicine think about herbs as a lifestyle.

> I would ask the American public to not look at plants as a drug in a bottle in the drugstore, but as an entire way of life. I am not talking about going into an ethnobotanical course in college. I am just talking about taking hikes out in the woods and looking at plants, noticing how beautiful they are. Maybe reading a book or two on the interrelationship of plants and humans, so that you have an overall vision of what plants really are about. A very small part of what they are about is medicine.

Simply put, living herbs as a lifestyle means that you interact with plants on a regular basis, that you value them as more than just medicine (in contrast to the botanical medicine paradigm), that you respect the plants' health and vitality in the way that you respect yourself and your own health. In this way, a more holistic model of living is built using herbs and herbal healing as the gateway.

Earth Connection Blossoming

An extension of "herbs as a lifestyle" is the deepening connection with Earth and nature. As holistic herbalists develop their practice, they necessarily become increasingly aware and tied to Earth. Ellen Evert Hopman (interview, 1999) illuminates this connection.

> Herbalism is learning how to use plants for medicine, for food, for crafts, having a deep relationship with nature, knowing what to gather at what season of the year. If you are an herbalist, you know, you have your own wheel of the year. There is a time when you get the mugwort. There is a time when you get the horse chestnuts; there is a time when you get the walnut hulls. There is time when you pick the hawthorn berries; there is the time when you get the roots; there is the time when you get the flowers. You are living with the green world in a very intimate relationship.

Those are your buddies. Those are your friends. Those are your counselors. You are using that knowledge for your own health and for other people.

In *The Male Herbal* (1991), James Green describes the process by which this connection develops in the herbalist. He names a reciprocal relationship between the herbalist and nature.

In the wake of civilization's current ecological dilemma, the practice of modern herbalism asks something of us at the same time that it gives to us. It asks us to reintegrate with our planet, Earth, and to reconnect compassionately with all of Earth's companion species; to take responsibility, responsibility for one's self, one's relationships, and one's surroundings. It asks us to relate to herbs not merely as material medicines but as allied planetary connections, vital parts of the whole of our lives that can help us revive a balanced relationship with a planet that we need to put back into order, thereby preventing most of our afflictions. It promises greater health, effectiveness, and abundance in return. (p. 9)

In living more closely with the Earth, holistic herbalism builds and expands on the "natural drug" framework of botanical medicine, and potentizes the effects of plant medicines.

Grassroots Organization and Emphasis on Independence

Herbalists feel that in contrast to the top-down, hierarchal nature of the biomedical model (and perhaps the botanical medicine model), holistic herbalism is a grassroots effort: medicine for the people, by the people. David Hoffmann (interview, 1998) describes it as an "expression of people following their hearts and then healing." He goes on to comment excitedly that in the beginning, "There were no leaders.... The herb revival started as total anarchic green rediscovery."

Teachers such as James Green make it a point to keep herbalism on that grassroots level by empowering people to become involved with their own health and ecology. Green (interview, 1999) calls it the "technology of independence." This involves people learning to grow their own food and medicines, make their own medicines, and provide basic, simple health care to their family and community, thereby reducing their dependence on the market economy. Green commented that herbalism is falling into the hands of large companies who are "marketing herbs like drugs" and taking the independence away from the people:

> People like myself are out talking to the public, saying, "Wait a minute—don't forget your independence!" I'm writing books now on how to make plant medicine, how to harvest herbs, how to grow them. I want to make sure people realize that herbalism is going full circle. If the herbal community is not careful, we'll end up being dependent on a marketplace again. That's my line right now; that's what the California School of Herbal Studies is about. I teach people to continue their independence. (James Green, interview, 1999)

Following the theme of independence and self-care, Michael McGuffin (interview, 1998), co-founder of McZand Herbals, defines holistic health and healing as "self-responsibility."

> The concept that I am most able to identify is something about taking care of yourself. This comes from my personal experience, my personal background, and my age and time and lifestyle choices. I thought it was interesting and a little in the face of the establishment to say that maybe if I were to learn enough about paying attention to my own body, I don't have to go to a professional. The professional need not be the best caretaker for the health of my body. Maybe I could play that role. That is clearly a holistic concept, self-responsibility.

Although a great deal of self-care is seen in the botanical medicine paradigm, the idea McGuffin expresses—that an individual can learn his or her body and therefore become a "doctor to the self"—is central to the more self-directed and grassroots model of holistic herbalism. Herbal practitioners mentioned repeatedly in interviews their desire for people to take responsibility for their own health and at times expressed their conviction to discontinue relationships with those who would not. In this way, the holistic herbalist ensures that basic herbal practice remains in the hands of people rather than becoming a professionally driven practice.

One of my biggest problems as an herbalist has been having people take responsibility for their healing. I find that when people come and consult with me, they want me to do what the doctor does. Take charge. Tell them what to do. And be responsible for their health. I won't do that. My job is to teach them how to take care of themselves. I don't use herbal medicine in a vacuum either. I am very much a holistic practitioner. I try to teach [people] to improve their diet, to deal with the stress in their [lives], to do different things. I really believe in [the] patients' responsibility for themselves. If someone doesn't do that, I don't want [him or her] for a client. I tell [the person] to go back to the medical doctors. (Steve Horne, interview, 1999)

I feel that the patient needs to work at least as hard as the doctor, if not harder. If I have a noncompliant patient, then I usually withdraw. If I am expected to devote a hundred percent of my intellect and my understanding to the care of a patient then I am expecting [him or her] to participate one hundred percent. It has happened—if not a lot, certainly frequently—that people have said to me, "Ryan, I can't devote my life to getting better." I have had to say, "Then I am sorry, I can't devote my expertise to your partial participation in this. Your life is at stake. I wish you would pay attention." Some of those people have died, sometimes relatively soon after seeing me, from if not preventable certainly ameliorative conditions. They would have been amenable to improvements in the qualities in their lives, I believe, if they had listened. One never knows. After they are dead you can say anything about it. So, I don't think that I accept responsibility for anything other than being as accurate and thorough as possible in what I say. It is the [individual] that actually does the healing. I am merely a vector for information and energy and glad to help out. (Ryan Drum, interview, 1999)

Inclusion of Both Scientific and Folkloric Ways of Knowing

Contrary to the accusations of biomedical and many botanical medicine practitioners (Angell and Kassirer, 1998), holistic herbalists do not deny the usefulness of scientific methods for testing herbs. Repeated comments were made throughout interview sessions that such testing would provide fascinating information to *corroborate* traditional evidence. Although herbalists feel that an understanding of biochemical mechanisms of plant medicines is not essential to the practice, they generally agree that it does enrich the practice.

> I actually love reading research, although it is always secondary to me to the herbal wisdom that gets handed down through the ages. A lot of the research is just saying the same thing that people already know, but it is presented in a different way. I have always found it fascinating to read the information. (Sharol Tilgner, interview, 1999)

> We are concerned that the paradigm in which herbalism operates will be preempted by the medical paradigm unless we do something to put our own philosophy forward and add enough Western science so that people feel comfortable and understand that we are competent. The key here is maintaining that holistic paradigm. If medical doctors take over herbs but don't learn herbalism then what we will have is the same thing that happened at the turn of the century. The Eclectic movement was supplanted by modern medicine. Herbalism went out of favor. For example, one of the issues that a lot of herbalists have is that as herbal healing moves into a medical model, what you have is standardized extracts where you concentrate certain constituents. People say this is more scientific. Well, that is what happened before. That is what ruined the Eclectic movement. In a lot of cases, we know from . . . experience that some of those standardized extracts don't work as well as the whole plant. That is a whole different way of thinking about it. (Steve Horne, interview, 1999)

The caveat generally appeared, however, that a new way of looking at the modern scientific information, a "softening" of science, is needed. Orest Pelechaty (interview, 1999), an herbalist and acupuncturist, spoke to this point:

> The scientific approach, which is to test hypotheses, has for a couple of centuries been tremendously skewed toward a left-brained, male-dominated, patriarchal, abstract, misogynistic construct. It has been polluted by that. It doesn't intrinsically

need to be that way. You can use logic and reason and experimentation to approach other phenomena that have been excluded a priori from this skewed, stunted view of reality. Science can indeed be expanded to include larger paradigms, nonmaterialist paradigms. There is that possibility. There could be a renaissance within science. It would be a big step to get the monied interest, the political interest, the corruption, the stranglehold of that from the scientific process, but I think that it is at least potentially possible. We can use a scientific approach, but a more holistic science. There is a hope for that.

In *Out of the Earth* (1991), master herbalist Simon Mills expands this thought:

We do not have to abandon the achievements of medical science to date. Rather we need a different way of organizing our knowledge, to bring it in line with ancient truths, and strive for a grand synthesis of the old and the new, a hybrid that vigorously does justice to both. (p. 11)

The main thing of value to most herbalists, however, is tradition. The use of plants as medicine predates all science—ancient or modern. Modern science may enrich the empirical evidence gained over the preceding generations, but it cannot supplant it. Further discussion of this point will be made in Chapter 3.

Body-Mind-Spirit Connection

I don't know what an average American thinks holistic means. Perhaps, they think holistic means "body, mind, spirit." That's good, but they need to understand that "body, mind, spirit" are not divided into three categories; holistic means that there's no division. The body is as sacred as soul, as sacred as emotion. We can treat everything as being sacred. Take anything as sacred; everything as sacred. When people get to the point where they can bring all of that together and start to really appreciate everything, even plastic.... That's holistic! (James Green, interview, 1999)

The key to holistic herbalism's understanding of this precept is the lack of divisions. It is not enough to recognize that the mind and the spirit have an effect on physical health and vice versa. One must understand that they are one and the same; each is a manifestation of a whole we cannot yet see. Once this is realized, it becomes impossible to separate each into its own category (David Hoffmann, interview, 1997).

Steve Horne (interview, 1999) spoke about the results of separating the being into parts.

> The Cartesian philosophy of separating the body and the spirit, of separating us into pieces, has created a schizophrenic culture. You cannot separate the whole. It ceases to exist. You can't look at the heart apart from the liver. You can't look at the disease of the person apart from who they are and the relationships that they are in.

Daniel Gagnon (interview, 1999) explains the range of different aspects to take into consideration when healing from a body-mind-spirit perspective.

> I look at certain aspects of health including things like rest, exercise, emotional support, having a base of friends and people to talk to, having goals, having a spiritual connection of some kind. I want to address all of that as I am addressing health.

Some herbalists who might be better termed vitalists subscribe to the view that the "vital force" or spirit is the life-giving, nourishing piece of the individual (Wood, 2000). They recognize the essential wholeness of body-mind-spirit, but lend more consideration to the spirit aspect. One holistic herbalist elaborates on the theory: "[If one] supports the vital force of the person . . . the health of the person . . . [his or her] body resolves things through [his or her] own abilities" (Peeka Trenkle, interview, 1999).

Kate Gilday (interview, 1999) suggests that through getting to know the person and the whole plants, an herbalist can find a single remedy to affect change in the whole individual—body-mind-spirit.

> There is no place that we can say, "Body over here, spirit over there." . . . When the human being—when the expression of that

human being is blocked, when the life force or the energy is blocked, you can figure out the right plant for that person by knowing that plant intimately and by allowing the person to tell his or her story.

In moving from botanical medicine to herbalism, not only has another aspect of the individual (spirit) been added for consideration, but the idea of body-mind-spirit has become much more central to the healing process as well. As we can see from the words of the herbalists, an individual looked at in his or her whole state is dynamic and capable of effectively initiating his or her own healing processes.

Treat the Person/Attention to the Individual

"Treat the person, not the disease" is an axiom of holistic herbalism. At its core, holistic herbalism is a highly individualistic healing art. Most herbals that provide formulas or recipes for ameliorating health issues usually mention this quality in the beginning, giving the reader permission, or rather compulsion, to change the formula to suit the needs of the individual.

In order for an individualized formula to be created, herbalists must spend a length of time with clients, allowing them to tell their story. Intimacy, Corinne Martin says (interview, 1999),

> is something that is really crucial and critical in terms of [herbalists'] sustainability as healers. [An herbalist sits] with a client for an hour, hour and a half, and [he or she is] intimate with that person. It is a whole cycle where every step of the circle of healing is intimate and tangible. . . . I think that . . . the potential is there for something to get lost that is really critical to herbalism, and I think to healing in general. I hear practitioners talk about it all the time. A doctor can see [tons of] people in one day, whereas I don't know any herbalist that can see many people in one day. I know that will change to some degree, and yet I am hoping that we can somehow maintain a kind of vision of the necessity for that kind of intimacy with the land, intimacy with the patient or client, intimacy with other practitioners. When the connections are that tangible and direct what we end up with is a lot more whole.

Simon Mills (1991) attributes part of the popularity of herbalism to this intimacy, citing the dramatic lack of it in modern medical practice.

> Modern doctors have relinquished the shamanistic or priestly role of addressing the patient's whole world and have sought a specialist job as a body mechanic. Alternative or complementary medicine has clearly claimed to provide more time for patients, a listening service, and then offered them imagery and models of their illness that they can relate to . . . their very popularity, at least as a salve for modern metropolitan neurosis, is a simple reflection of the poverty of the modern medical story. (p. 21)

For Mills, the attention shown to the client is an aspect of healing. By following the body-mind-spirit approach, the consultation allows the client a release of sorts and an affirmation that his or her thoughts and feelings are appropriate and important.

The uniqueness of holistic herbalism as opposed to botanical medicine is that the individual is allowed to be just that—an individual. Botanical medicine, as an extension of biomedical science using plant remedies rather than pharmaceuticals, seeks "standardized patients . . . for standardized medicines" (Ryan Drum, interview, 1999).

Having described the differences and connections between botanical medicine and holistic herbalism in some detail, let us return momentarily to the idea of the continuum of healing paradigms or systems. In speaking of what is beyond holistic herbalism, drifting further from the biomedical model, it is notable to mention homeopathy, flower essence therapy, and plant spirit medicine in particular. These three modalities represent a concentration on energetics and subtle influences on health and wellness (Wood, 1997; Cowan, 1995). They seek to affect the physical body, using methods that employ the spiritual and emotional realm predominantly. These healing modalities fit into the linear model at various points around the Davis-Floyd and St. John (1998) paradigms and those of American herbalism discussed previously, and it may be that some need to be situated on their own continuum overlaid on the current model. Such explication is beyond the scope of this text, however.

This book describes American herbalism. The objective of laying out the model and discussing points along a continuum is to illuminate two general threads of herbal healing in America: botanical medicine and holistic herbalism. In every day vernacular, the term *herbalism* is often applied to the use of herbs as medicine regardless of the specific approach being utilized. Furthermore, *holistic* refers to any kind of alternative or complementary medical practice. However, when these terms are united, *holistic herbalism,* they denote a specific practice involving plant medicine. In its holistic form, herbalism is a way of life, a way of being, of perceiving, and of interacting with the world. It suggests an attachment to plants that is greater than a capsule of ground plant material taken once a day. It is more than the "cure" for depression or memory loss, although that is a part of it. Holistic herbalism reaches into the body-mind-spirit and gives its proponents a sense of connection to beauty, to rhythm, to sensuality.

All methods and approaches are valid and necessary to enrich and enliven the American system of healing. In any field of interest, there are those who are curious and experimental and those who are completely swept away by their practice. The people interviewed for this study fall into the latter category. They live, eat, sleep, and breathe herbalism. Their lives are dedicated to educating about herbs, natural medicine, health, and alternatives/complements to biomedical treatment.

One of the initial questions in this chapter was what the connection is between James Green's sensory description of herbalism and the projected billions of dollars spent annually on herbal products. Simply put, the $3 to $4 billion industry greatly benefits from and drives the botanical medicine paradigm—the application of specific herbs for specific healing. The American herbal renaissance is most directly taking place in the natural food stores and corner drugstores across the country. However, as the mass of people willing to experiment with herbs increases, some have to be completely swept away by it. Those are the people who consistently fill up herbal classes and conferences presented by many of the narrators of this story. The teachers, clinicians, gardeners, wild crafters, and business people who were interviewed demonstrate the possibility of a life that truly values the self, the Earth, the community, and the relationship between the three. Holistic herbalism demonstrates that these ideas are not an ideal from a far off, traditional culture; the lifestyle is a possibility even within the context of the modern world.

Holistic herbalism and its philosophy must be considered a work in progress, a process more than an entity. It has laid out an ideal; it has helped to push the vision of the mainstream public further toward an Earth-centered holism in some ways, but it must continue to strive for stronger, lasting bonds between the self, Earth, and community.

This study seeks to paint a picture of herbalism in America at the dawn of the twenty-first century. Through dialogue around four critical issues—(1) the debate over scientific exploration versus folkloric approaches to healing, (2) questions of licensure and professionalization for herbalists, (3) the plight of endangered medicinals and the ecology of herbal healing, and (4) the necessity for a deeply rooted spirituality within American herbalism—it is my hope that the reader may gain further insight into the ideals, realities, struggles, and areas of lacking within contemporary American herbalism. With an open mind and heart, read on.

Chapter 3

Recycling Science and Grasping the Ungraspable: Cartesian Thought and Folklore in Herbal Practice

To begin to talk about science without intending to dedicate an entire volume to its many subtleties and variations is perhaps folly. Nevertheless, the interplay between science and American herbalism is one of the hottest and most dangerous topics going and to leave it out of a synthesis of the field would render this discussion incomplete. As I mentioned in characterizing botanical medicine and holistic herbalism, the former depends upon the language of conventional biomedicine a great deal, and the latter struggles with being inclusive, allowing science and folk tradition to exist simultaneously. Each paradigm within American herbalism respects the idea of science as knowing by observing and documenting, but there is a strong objection and desire to leave behind some of the more negative qualities of an industrialized science: reliance on a mechanistic, reductionist framework rather than on a more dynamic, interconnected web (Capra, 1983); human as dominator with regard to ecology (Berry, 1988); and the move to squash or co-opt that which is competition (Whorton, 2002), to name a few. All herbalists have their position on the use, misuse, inclusion, or exclusion of science and hold onto it stridently, at times foregoing alliances that might serve to advance the community as a whole. The debate over science versus folklore—experimental versus experiential learning—has the strong potential to divide the herbal community into two irreconcilable halves.

But what is meant by science? Is it simply work done in a lab? Is it experimentation? Certainly alchemists and herbalists have done this for thousands of years. What is folkloric evidence, if not the result of experimentation and observation? Many herbalists do not object to this information, but, on the contrary, base their practice on it.

Science refers to a wide range of approaches. A distinction must be made, however, between the branches of modern science that utilize an isolationist framework in which entities are viewed as a closed system, showing little significant interaction with their surroundings, and those that subscribe to a relational architecture in which the entity is an open system constantly influenced by the surroundings (Capra, 1996). The science that holistic herbalists reject is based on the fallacious theory that the experimenter can somehow separate himself or herself from the subject, removing the influence of the surroundings (the experimenter and experimental procedure), and thereby produce objective results. Modern biomedicine is based on this idea, which is descended from the work of Aristotle and Descartes (Illich, 1977).

There are branches of science that acknowledge the relationship between entities. Among these, quantum physics has come to terms with the understanding that the observer changes the subject and vice versa. The change or influence can be as simple as the experimenter bringing his or her bias to the design of an experiment, or as esoteric as the electrical energy required for measurement altering the electrical state of the subject, thus preventing one from observing the "normal" pattern of the entity. The latter example gave rise to the Heisenberg uncertainty principle (discussed in greater depth later in the chapter).

Although holistic herbalists value biochemical analyses and suggested mechanisms of action of plant medicines, they are cautious about falling headlong into a biomedical science whose basic philosophy runs counter to the herbalists' idea of the relational dynamics between entities. The holistic herbalist objects to a compartmentalized mode of healing in which plant constituents or patients are viewed as isolated entities. In fact, in the case of the patient, the nature of his or her many relationships to family, friends, the environment, his or her spiritual self, and indeed to the healer, is of core importance to the holistic healing process.

To more fully understand the interplay of these two braches of science—biomedicine and quantum physics—the continuum presented in Chapter 2 may be elaborated upon. Due to certain similarities in thought, biomedicine would overlay the botanical medicine and technocratic paradigms, and, on the far end, over holistic herbalism and its holistic medical counterpart would be a scientific discipline such as quantum physics, which understands the universe as a set of interwoven relationships and admits to the folly of "objective," linear science. Figure 3.1 demonstrates this idea. It is with the more holistic of the sciences that American herbalism may find healthy relationship.

First, however, we must understand how American herbalism reacts to the presence of biomedical science within the domain of plant medicines as well as the function of folklore in the practice of herbalism.

HOW DOES MODERN BIOMEDICINE ENHANCE THE PRACTICE OF HERBALISM?

"[T]he path of the herbalist may . . . encompass the laboratory with the identification and extraction of certain vital biochemical agents found in herbs" (Tierra, 1990, p. xxiv). This may not be an overwhelmingly popular sentiment within the herbal community, but most herbalists are coming to terms with the use of science (and biomedicine specifically) in American herbalism. As the two interact at greater length, it will become possible for the dialogue between them to grow deeper. Even at this preliminary stage, biomedicine can be said to enhance the use of herbs in a variety of ways.

First and foremost, an understanding of medical science gives an herbalist a familiarity with physiological pathways in the body. As American herbalism becomes increasingly popular, herbal clinicians are frequently confronted with patients concurrently seeing physicians for serious ailments. These patients come to herbal consulta-

Biomedical Science		Quantum Physics
Botanical Medicine		Holistic Herbalism
Technocratic	Humanistic	Holistic

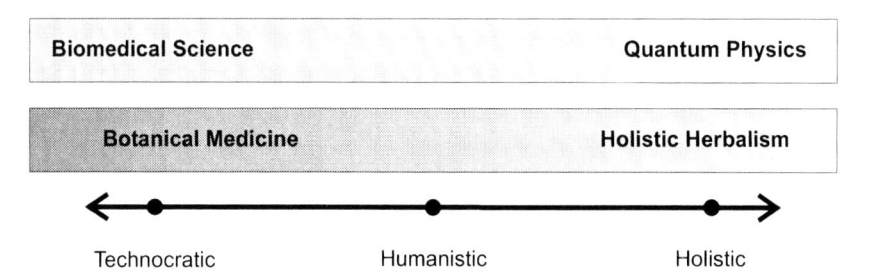

FIGURE 3.1. Biomedicine and Physics Comparison Continuum (*Source:* From Davis-Floyd, Robbie and Gloria St. John, *From Doctor to Healer: The Transformative Journey.* Copyright © 1998 by Robbie Davis-Floyd and Gloria St. John. Reprinted by permission of Rutgers University Press.)

tions with a medical explanation given by their physicians. It is imperative in cases such as this that the herbalist comprehend the language of life science and the related descriptive pathologies presented in order to enrich the complementary treatment of the patient as well as to help gain the patient's confidence in the herbalist's skill. David Milbradt spoke to this point (interview, 1998):

> Simon Mills, in his book, *The Essential Book of Herbal Medicine,* says that one of the differences between the doctors and the herbalists is that the herbalists have better stories. This is important because if you can't help a person to understand how [he or she] can move from disease toward health, if you can't give a clear picture of how that can happen, the patient may not be able to visualize the step he [or she] needs to take to get there. The scientific story of how the body is making that step can be every bit as important and operable as another story. In some ways science gives us a tool there to help explain to the person how [he or she] can move from disease toward health. In that sense, it is very useful.

Understanding the anatomy and physiology of the body also allows the herbalist to communicate with the physician either directly or through the patient. A coherent conversation or relay of information will allow the herbalist to be more credible to the physician as well as more helpful to the patient who seeks alternative options from the medical doctor. It must be remembered also that just as the herbalist must adapt to the language of medical science, so too must the physician modulate to the theory and jargon of the herbalist. For example, the physician must develop an understanding of the importance of "system tonics," a concept that has very different meanings to traditional herbalists and modern scientists.

Corinne Martin (interview, 1999) related a secondhand account of a situation that could have been enriched by the use of herbs but was not because of the herbalist's lack of scientific knowledge.

> I remember speaking to an herbalist once who said that the hospital called her and said this person had such and such and was in the emergency room, and she refused drugs. What herbs would they suggest? The woman didn't know the words that they were saying. And you know there are a million medical

words that you cannot know, but to at least try and familiarize yourself with that language I think is really important.

Although herbalism has its own belief system based on an idealized view of the body in health and balance with Earth, the biomedical model must also be given its due credit, as it offers an interesting and intricate model of the biology of the human body. If the stated goal of herbalism is to treat individuals holistically, it seems that the herbal healer (as well as the physician) needs to utilize all available information/models. David Hoffmann (1992) writes of this synergy: "From a more positive perspective, cooperation can lead to synergistic support, with the whole of any treatment program being more than the sum of its parts" (p. 252).

In thinking about herbal medicines as healing agents, clinical trials are useful in exploring the healing potential of individual herbs and their constituents. With this knowledge the herbalist can be more precise in his or her treatment protocols. For instance, when a nervine sedative is indicated, an understanding that valerian reduces sleep latency while California poppy lengthens REM sleep allows the herbalist to more directly affect the place of healing (Low Dog, 1997). Why give valerian to someone who has issues with early morning waking when California poppy would more effectively treat the condition? Such knowledge is available through scientific study that examines the active constituents and mechanisms of herbs.

Many herbalists acknowledge the potential usefulness of biomedical science, but they do so with an air of caution. Misgivings with the methods of scientific inquiry arose in nearly every conversation concerning the topic. Many herbalists did not trust the standardization and isolation of herbs and constituents that modern scientific inquiry requires. Others were turned off by what they viewed as the egotism and elitist nature of the discipline. David Hoffmann (interview, 1998) noted that,

> The only people that should get into the science are people who know the science enough so they can step out of the science and ignore it. If you start from the place of really being embraced by Gaia, and you are attracted to science, then it is worth doing the two.

James Green (interview, 1999) echoed this sentiment.

> There could be a person who loves to garden and still loves to read about sesquiterpenes and flavonoids. It trips them out. They like that point of view. If you can have that kind of person who enjoys organic chemistry and also likes to lay in the red clover and dream about plants and devas, then that's great. The problem is when someone comes along and says, "Oh, unless you understand the pseudoscience of it, you can't be a good herbalist. Unless the product has been standardized to a certain active ingredient, it can't be any good." That's where I have my problems. We get this quiet dominance that says, "Well, if you don't understand the language of science, then you're not good" or "it's not a good product." Students who come to my school think they have to understand organic chemistry, and they don't. If they choose to learn it as a component of their herbal studies, that's great. It is not required, however. You don't have to do a weight-to-volume tincture to create a good tincture. You can work with the folk methods and make a very good tincture. Healing with plants has been around much longer than the weight-to-volume tincture.

In an article written by scientifically trained professionals with an interest in complementary medicine (Eskinazi and Muehsam, 2000), it was noted that subsuming traditional values to science in order to gain acceptance may compromise the paradigm-shifting ability of complementary medicine. The authors stated that

> overzealous attempts to meet these criteria may, paradoxically, be counterproductive. Rigidly following orthodox research practices to prove the effectiveness of a therapy and proposing scientifically plausible mechanisms before the phenomena under investigation have been adequately explored may hamper the fair evaluation of alternative practices. Also, attempts to formulate theoretical models for practices that may be effective through mechanisms that are ill understood within the boundaries of current biomedical understanding can result in unnecessary narrowing of research scope. (p. 49)

The comments of the majority of the study participants acknowledge the validity of this statement and point to a balance and unity that must be found between science and folk tradition. The field of

American herbalism is in many ways a direct reaction to the domination of biomedicine and scientific technology. Physician Frederick Stenn (cited in Whorton, 2002) summarizes this alienation from biomedicine.

> Most physicians have lost the pearl that was once an intimate part of medicine—humanism. Machinery, efficiency, precision have driven from the heart warmth, compassion, sympathy, and concern for the individual. Medicine is now an icy science; its charms belong to another age. The dying man can get little comfort from the mechanical doctor. (p. 219)

If one understands the contemporary American herbal renaissance as beginning with the search for independence from society and biomedicine that blossomed during the "back to the land" movement of the late 1960s and early 1970s, then it is well within reason that the herbal community might be skeptical of scientific inquiry into their field. However, the hope for a new and better synthesis of health care, which merges scientific and folkloric methods of inquiry, is prominent. As James Green (interview, 1999) poignantly said,

> I think if a person is really holistic [he or she will] allow things to be. [He or she] will not want to gang up against a certain idea, or against a certain group because [he or she doesn't] agree. I really feel that the most enriching space to be in is that of compromise. If there are two opposing ideas out there, or two individuals with different ideas, I think that the creative compromise that comes from the meeting of the ideas is often the best. I feel that the pharmaceutical community and the herbal community are going to come up with something far beyond both of them in time. If we can get the sense of opposition, us against them, out of there and just sit and honor each other as individuals, we can see how we can work together as opposed to only for ourselves.

WHAT DOES HERBALISM DERIVE FROM FOLKLORE?

Why is folklore so important to American herbalists, especially those practicing holistic herbalism? What information about phytomedicines and their applications is essential to herbal healing?

Dioscorides, in writing about hawthorn berries, said that they "regulate the fluxes of blood" (cited in Gerard, 1975, p. 1328). Today it is used by herbalists to moderate blood pressure and tone heart muscle. John Gerard, sixteenth-century botanist and collector of herbal lore, says, "The juice of licorice is profitable against the heate of the stomacke, and the mouth" (cited in Gerard, 1975, p. 1302). Indeed, the twentieth-century herbal practitioner loves licorice for its "anti-inflammatory and anti-arthritic effect similar to hydrocortisone" (Mills, 1991, p. 506), with its affinity for the digestive tract.

The wealth of knowledge provided to herbalists by the age-old data-gathering service, tradition and folklore, is immense. Although some of the information refers to outdated conditions and treatments no longer applicable to modern life, and much of it is so buried in the language of the day as to be unintelligible to the modern reader, if an herbalist takes the time to sift through the written volumes or oral testimony as it has been handed down, he or she may uncover poignant and useful advice for the employment of herbal medicines in contemporary practice. Herbalists in this study repeatedly extolled the virtue of folklore's long history. They especially utilized a commonsense approach; if the plant did not do what it was said to do, nobody would remember.

"The whole plant worked for five hundred years. That is how we know about it in the first place. It never was standardized, and it worked" (Daniel Gagnon, interview, 1999). The modern practice of herbalism relies heavily on the wisdom of tradition. In this way, herbalists honor the past and those who paved the way for them. Steve Horne (interview, 1999) touched on this idea:

> There is this idea that we have in our society that our civilization is smart, and everybody that ever lived before was stupid. It isn't true. While you do find funny stuff in herbal folklore, what you also find is that when people have used the plants traditionally, you can usually depend on that information being fairly accurate. Let's take yarrow, for example. Yarrow was used to treat battle wounds on three different continents for over two thousand years. To dismiss that and say, "Oh, that was just folklore. That was just imagination. All of those soldiers who had yarrow stuffed in their wounds, that was just a placebo" . . . is awfully arrogant. I have stuffed yarrow in a wound and watched it do exactly what the tradition says it does.

After trying the therapy as suggested in folklore and witnessing the same or similar results as the tradition dictates, herbalists tend to pass the information along to other herbalists. Corinne Martin (interview, 1999) says she does not feel comfortable using a therapy unless she can "talk to people who have used plants and find out what their experiences have been."

It is through this communication that community is built and maintained. Again, this highlights the herbalists' reliance on relationship—here, the practitioner-practitioner relationship. Herbalists need the network of other herbalists and the sharing of information in order to proceed. Herbal conferences such as the International Herb Symposium or the Green Nations Gathering are held for this purpose, implicitly or explicitly. They provide an opportunity to share case studies, successes, and failures; learn new information; and recognize other people who are involved in the same pursuit. In this way, modern folklore (the experiential learning of contemporary practitioners) and its use may be seen as a starting point for community building.

Perhaps most important in a discussion of folklore and tradition in holistic herbalism is the mysterious and intangible side of health and healing that it holds so dear. In *The Dream of the Earth,* Thomas Berry (1988) wrote that we as humans desire and feed on mystery. In order to truly enter into the next phase of life, the ecological age, he postulated that we must recapture the mystery and add mystique to our language of exploration. James Green spoke about this in terms of herbalism at the inaugural meeting of the American Herbalists Guild in 1990. In outlining what was essential about folkloric herbalism, he said that it is the unknown, unknowable mysteriousness of the folkloric tradition that has allowed it to persist through the ages.

The process of making herbal medicines in the paradigm of holistic herbalism utilizes the mystery that Green discusses. Each step in the process must be imbued with ritual acts that gather the spiritual and intuitive qualities. In *Sacred Plant Medicine,* Stephen Buhner (1996), a holistic herbalist, writes eloquently about the necessity of ritual in human life:

> It has long been known among indigenous cultures that when people forget their place in the web of life without periodically renewing connection with the sacred, illness and disharmony follows. . . . This propensity to lose connection is probably a

normal and natural part of our human makeup. Connection must be renewed through ceremony over and over again. (p. 104)

It is through ritual and conscious intent that holistic herbalists harness the mysterious and ephemeral side of life, the nonlinear, nonquantifiable quality that they purport strengthens their practice. Kate Gilday (interview, 1999), in talking about the commercialization of herbal medicines, points to many essential aspects of plant medicine that are not given due consideration in the botanical medicine context (and certainly not in the biomedical arena) and gives simple examples of intention in medicine making. Points about intention in medicine making and healing are part of the mystery of holistic herbalism that are discounted in botanical medicine and biomedicine because they cannot be quantified. If one lacks the relational understanding of life, the concept of the web of life, it becomes difficult to acknowledge that the harvester affects the plant, the plant affects the medicine maker, who in turn affects the medicine, and so down the line. An isolationist view of a plant as a container for certain chemicals does not allow for the interaction and transference of energy from medicine maker to plant to patient.

> Who gathered the plant? Did they gather it with respect? Did they just take a big mower and mow it down and pop it into a bag? The tincture maker? Did they shake [the tincture] daily? Did they say prayers when they were making their medicine? Is it just in a lab where everyone is isolated from the plant material? This is where the idea of supporting your local herbalist comes in very strongly. That makes a difference. (Kate Gilday, interview, 1999)

Annie McCleary, a bioregional herbalist in New England, elaborates on the role of spirituality and intention in medicine making.

> Herbal medicine involves much more than chemical constituents. It is not a mechanical process—simply grow a plant, chop it up, put it in—*no*. Herbal medicine making is full of intention, a constant honoring of the plant spirits. So by the time this little bottle gets to someone who needs it, this is charged; *this is charged.* A fellow came up to my booth at a local food and health fair. He told me that he uses my echinacea extract. Some-

times he ingests the extract, and sometimes he simply holds the bottle and feels a change in his vibrational energy. (interview, 1999)

From these herbalists, we learn that relationship, a dynamic involvement of herbalist and herb, and herbalist and client or patient, is essential to the healing practice of holistic herbalism. Holistic herbalism is not an objective practice; it is an interplay. Interplay, relationship, and involvement all connote a certain variability. Things are always different even if only slightly. Nothing can be repeated. Holistic herbalism invokes and even treasures the subtlety of relationship. Working with such subtlety and variability necessitates flexibility and a significant level of trust, which the holistic herbalist gains through study of tradition, respect for an elder's experience and wisdom, and a deep understanding of the plants and their affinities. The herbalist witnesses the plant world as a community of individuals who have diverse personalities. The holistic herbalist understands the materia medica, the range of plant remedies, as friends that you can never predict precisely, but as you know them better you are able to generate more consistent results.

Fritjof Capra (1983) describes medical science's discomfort with this variability.

> In modern scientific terms we could say that the healing process represents the coordinated response of the integrated organism to stressful environmental influences. This view of healing implies a number of concepts that transcend the Cartesian division and cannot be formulated adequately within the framework of current medical science. Because of this, biomedical researchers tend to disregard the practices of folk healers and are reluctant to admit their effectiveness. (p. 125)

The number of variables and the subtle aspects involved with holistic herbalism when it is practiced in its whole state is unacceptable to medical science. As holistic herbalism reaches beyond the biomedical paradigm, herbalists must take heed of Capra's words. Capra states that biomedicine has formed its opinion of folk healing out of a lack of knowledge of folk tradition and how it seeks to heal. It is, in part, an opinion forged out of ignorance. This point is significant as a warning to herbalists to refrain from doing the same, but rather open up to the benefits of biomedicine and explore more fully the shortcomings.

OBSERVATION AND CHANGE: THE LINK BETWEEN SCIENCE AND FOLKLORE

Having outlined the enriching aspects of folkloric and scientific inquiry, let's attempt to understand how the integration of the two may occur at a deep and meaningful level. If one closely examines both herbalism and science, one can see a small door between the two. That door, in my observation, is the Heisenberg uncertainty principle, a scientific observation with holistic underpinnings.

The new physics, or quantum physics, which has developed over the past half-century, concentrates on the smallest particles of existence—protons, neutrons, electrons, and their subsets, quarks—and attempts to observe and define their behavior. Early on in the development of the field, researchers ran into a problem, however. It seemed that as one attempted to pinpoint the position of these subatomic particles, the uncertainty in the particles' momentum changed and vice versa. Physicist Henry Stapp says, "An elementary particle is not an independently existing unanalyzable entity. It is, in essence, a set of relationships that reach outward to other things" (cited in Capra, 1983, p. 81).

Position and momentum are part of what identifies a particle as unique. The more certain an external observer is of the one quality, the more uncertain the other becomes. This essentially changes the identity of the particle for the researcher. In this realization, quantum physics discovered that the observer has an effect on the observed. There is no such thing as objectivity if one always alters the situation by investigating it. The Heisenberg uncertainty principle discusses this subjective relationship with relevance to quantum physics, stating that we cannot precisely measure anything due to the wide range of variables and the interlinked, systemic relationship of such variables. Frijof Capra (1975) describes the impact of this axiom saying, "The fundamental importance of the uncertainty principle is that it expresses the limitations of our classical concepts in a precise mathematical form. . . . The subatomic world appears as a web of relations between various parts of a unified whole" (p. 75).

Although relationship and variability confound classical models of physics, quantum mechanics can calculate an answer based on probability. Probability in the scientific world is a combination of guesswork and rigorous mathematical models. Heisenberg himself describes the results of this realization: "The world thus appears as a complicated tissue of events, in which connections of different kinds alternate or overlap or combine and thereby determine the texture of the whole" (cited in Capra, 1983, p. 81). This principle along with the work of many quantum physicists essentially annihilated the Cartesian model of reality—a reality built of pieces (like a car) that can be observed in isolated experiments—in all fields of science except the biomedical arena. Indeed, quantum physics has scientifically proven that the universe is a set of relationships which may not be examined except with regard to each other.

As mentioned, the use of herbs is steeped in relationship. In the practice of botanical medicine, those relationships may be thought of as between the patient and the herbalist or the herb and the herbalist (in a few cases). The relationship, however, remains one of objectivity more than interaction. In the practice of holistic herbalism, however, the terrain becomes much more dynamic. The clinical practice of holistic herbalism involves an expressive three-way relationship between the herb, herbalist, and client.

Each set of relationships is two-way as well. That is, each entity provides the other with knowledge and understanding of the situation at hand. Note the double-headed arrows in Figure 3.2, which symbolize the reciprocity of the relationship. For example, the client tells his or her story and states the reason for the visit. The herbalist in turn allows the client to tell the story (which may be thought of as healing through catharsis) and provides constructive suggestions. There is *interaction* between the two. Each participant tells a story which allows the healing process to begin.

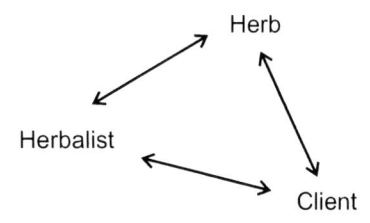

FIGURE 3.2. Holistic Herbalism Three-Way Relationship Model

In looking at the relationship between the herb and the herbalist, one must consider the nature of the herb or the energetic signature of the plant. In the same way that each human being is unique, and has his or her own energy pattern, so too does each plant. For example, when a hypnotic is indicated, some herbalists describe a certain herb from this classification "calling" to them. Perhaps the plant's energetics match precisely with the patient's. Other herbalists take this notion of energetic signatures a step further in saying that one plant in a patch of a single species "calls" to them. Kate Gilday described this phenomenon around medicine making:

> When we first decided to make these particular flower essences, the at-risk essences, it was more that we were coming into communion with the at-risk plants and we felt that they were asking us to do this, the plants themselves or Nature, herself. (interview, 1999)

These statements are often passed off by nonherbalists as figurative or metaphorical statements. Until recently, there was no way to account for this phenomenon in a manner acceptable to the scientific community. Recently, however, Jeremy Narby (1998), a Swiss-born anthropologist, attempted to formulate a scientific explanation for this situation. He was perplexed by the claims of Amazonian shamans that the plants of the rain forest were the source of their wide-ranging and amazingly precise botanical knowledge, that the plants themselves told the shamans what plants were to be used for specific purposes. After a thorough investigation of current research, Narby breaks the puzzle down to its smallest pieces, DNA. He theorizes that the DNA of plants communicates through the employment of photon emissions. Shamans, through consciousness-altering botanical preparations, adjust their consciousness and therefore sensitize themselves to these emissions. Although Narby's work is not based on studies specifically designed to investigate this issue, he adequately connects the dots of the existing scientific literature to create an interesting theory for consideration by the skeptical mind. If one can accept this explanation, the relationship between the herb and the herbalist may also be seen as twofold.

The two-way aspect of the herb-herbalist relationship will be for many readers the more difficult to accept, but I personally find the herb-client interplay harder to wrap my mind around. If one examines the physiological nature of the relationship perhaps it appears simple. The material herb is given to the client (→). The client takes the herb and hopefully benefits (←). We could extend this to say that the client is so pleased with the help of the herb that he or she plants the herb on a windowsill or in the backyard. Such an action straddles the spiritual and material worlds. The meaning given to this act depends on the intention of the individual. To truly achieve a spiritual or simply a nonmaterial "conversation" between the herb and the client is perhaps the most difficult job of the herbalist. This is a key area for the herbal community to examine as the field becomes increasingly popular. With the solidification of this relationship, the full potential of herbalism may be realized and herbalism can help to heal not only individuals but also assist in the creation of a more eco-spiritually aware society. The improvement will come slowly as the herbal community forges deeper spiritual bonds itself and as the general public becomes more accepting of such relationships.

As we examine holistic herbalism in more depth, we come to understand that not only are there three players in the relationship triangle—herb, herbalist, client—but that each relationship is two-way. Therefore, herbalism effectively embodies six links or sub-relationships. Compare this to most conventional biomedical alternatives that do not allow relationship to become part of the discussion. Biomedical, Cartesian science is rooted in the thinking that objective observation and static nature are central to existence. In discussing the use of herbalism in psychology, David Frawley (1992) writes about these intricacies:

> Such subtle realities cannot be described in gross physical or biochemical terms . . . otherwise they themselves would only be physical and not beyond it. Hence to approach the subtle qualities of herbs may require a different view of reality than that afforded by modern science. (p. 145)

In its strictest, limited form, the purveyors of biomedicine are not able to acknowledge the subtleties of herbalism and the vast number of possibilities for change and influence.

How can it be that one branch of science (quantum physics) can understand a relationship-based worldview and value it as the core of its research while another (biomedicine) cannot? How can biomedical science look into the face of herbalism and purport that it is quackery although the basis for so many biomedicines is plant medicine?

In reading Martin Prechtel's *Secrets of the Talking Jaguar* (1998), I have uncovered a possible answer to these questions. Prechtel discusses the Mayan belief that life is essentially about remembering. The Mayans believe that one is born through four layers of reality where one's entire being is pieced together and solidified before reaching this world, the fifth layer. At birth, one begins to quickly forget about the layers of creation, which are truly layers of the self. The cultural mission of life within the tribe is to remember the four previous layers and therefore fully realize one's self.

Perhaps medical science is doing just that. At one time, healing was about relationships, spirituality, and intuition. It straddled the fence between religion/spirituality and natural science. Modern medicine is essentially a child of herbalism as herbalism was the dominant modality of healing for thousands of years. With the advent of "objective" science and the quick development of technology, medical science drifted further away from its roots in botany and spiritual healing. Is it now on its return? Leslie Gardner (interview, 1999) articulated this pendulum effect:

> There are cycles to everything.... We have cycled ourselves so far away from plants, and now we are cycling back. It is sort of a natural progression. When you get too far away from something, you have to go back.... We have been in a patriarchy for so long that now we are cycling toward a matriarchy, slowly moving back to that over many thousands of years.

Perhaps the realizations of quantum physics are precursors of things to come. Perhaps quantum physics has already rediscovered some layers of its being and is more fully integrated into the cosmos in the Mayan sense. In remembering its true nature and taking cues from its ancestry, biomedical science may be able to leap into another dimension of applicability. Perhaps part of the job of herbalism is to aid in that remembering. To do so, herbalism must remain true to its present values and push farther into the realm of the spiritual and mysterious. It must remain open enough to converse with medical science and thus be a relevant fixture in the world of health, ecological renewal, and lifestyle care.

Chapter 4

Free-Form Goes Mainstream: The Debate over Licensure and Professionalization

In any discussion of critical issues affecting American herbalism, licensure is a necessary topic. Although no system of licensure exists within the herbal community, the American Herbalists Guild (AHG), a national organization of clinical herbalists that works to promote excellence in herbal practice and serves as a networking base for clinical herbalists, is working on developing a certification process for professional members that would be nationally recognized. Currently, three levels of membership are recognized in the AHG: professional, general (for those working toward attaining professional membership), and student. The present standards for professional membership into the AHG are four years of training in the clinical use of herbs and/or equivalent clinical experience, completion of a lengthy application, and submission of three letters of reference and three case studies. A personal and professional biography must also be presented. The applications are peer reviewed by a board of professional members and, upon acceptance, candidates agree to uphold an ethical practice as outlined by the group. Professional members may use the initials AHG after their name to denote their allegiance to the group (Romm, 2002).

Under the proposed national certification program, professional members would be known as registered herbalists and could prepare for a national examination to gain the title of certified clincial herbalist. The national exam is currently under construction. In the future, an entry-level exam will be given in order to apply for registered herbalist status (Romm, 2002).

At present, the AHG is the only national organization that discriminates between levels of herbal practice. Although the professional status offered by the AHG is not legally binding, the developers of the new certification process hope that it will bring rigor and excellence to the field as well as protect American herbalism from having other professional groups develop standards for herbal practice (Romm, 2002).

It is important to note that although many herbalists object to any sort of standardization of practice, a major difference exists between certification and licensure. Certification is a voluntary action taken by professionals wanting to distinguish their level of practice, while licensure is a mandatory process that makes it illegal to practice without the license. The AHG supports certification and says, "The AHG will oppose all legislative efforts that attempt to limit the practice of herbalism to any one class of people, professionals or organizations" (Tierra, 2002, p. 2).

For the AHG, certification is essential for the health and independence of the American herbal community; for many other herbalists across the country, however, licensure or certification means certain death for the folkloric practice of herbalism. The debate over licensure has herbalists split into two camps without much in-between space. The discussion is ongoing, and I will briefly outline the findings of my interviews in this chapter. I will begin by presenting some of the reasons why herbalists support licensure, and then explore the reservations that many herbalists have with the idea. It must be noted that interviews were conducted throughout 1998 and 1999 when the idea of AHG national certification was in its nascent stage. Therefore, the narrators' responses reflect general feelings about licensure rather than the particulars of the AHG plan previously outlined.

A CASE FOR LICENSURE

Public safety is usually the first reason proponents give for the support of a licensing procedure. Aviva Romm, director of education and certification for the AHG, prepared the organization's notes to the White House Commission on CAM (complementary and alternative medicine). She wrote that while

> no certification process can guarantee the integrity or quality of individual practitioners . . . it does provide a baseline assurance

of quality and accountability. Thus patients or other professionals who hire Certified Clinical Herbalists will have the guarantee that those individuals went through a rigorous assessment process and met stringent standards before bearing our credential. (2002, p. 4)

As herbalism is relatively new to the United States, those wanting to use herbs for health may lack a cultural context, a cultural understanding, of how to use herbs properly or select herbal guidance from a practitioner. Many people turn to the clinical practitioner for advice on how to treat conditions using herbal medicines, leaving the treatment of their own health issues completely in the hands of the practitioner. Without a systematic way to certify that a given herbalist has the requisite knowledge to provide advice on the health issues at hand, an unknown danger is before the consumer. Does the herbalist understand the condition? Is the herbalist using appropriate and effective herbs in the treatment? Without a professional standard, the consumer has little way of knowing.

In a traditional society, each village or tribe has its own healers who use herbs. The village knows which healers are competent and which are not. If the healer does not successfully treat those seeking assistance, the healer most likely loses his or her clientele. The village is the "licensing system."

Since the United States lacks this village structure, it is difficult for the client to know whether an herbalist is competent. David Hoffmann spoke of the need for protection of the public as herbalism increases in popularity.

> The reason we need to do [licensure] is for the safety of patients. It is not to further our egos or money. There are some really dangerous ideas out there—bad medicine, herbs used inappropriately. We herbalists shouldn't be doing that. One way of protecting the general public, who are coming from herbal drought, is to police ourselves. (interview, 1998)

In initiating a certification procedure, the AHG has also focused on raising the educational standards for students of herbalism. The education and certification committee is beginning a school assessment program (Romm, 2002). Those graduating from any approved herbal "college" across the country will have studied a standard curriculum of subjects and be prepared to pass licensing exams and/or review boards. Through higher educational standards, a canon of herbalism, a specified course of knowledge, may be codified, which some herbalists believe will provide the general public with better, more consistent results. K. P. Khalsa hopes to show the American public that herbalism is dynamic and powerful. He goes on to speak about the benefits of a codified herbal system with higher educational standards:

> The advantage of a more formalized system is that people are going to get much better clinical results. There will be much less misuse of popular herbs. Here, probably ninety percent of the echinacea is taken improperly, for the wrong uses, in the wrong dose; the quality is bad; and it is clear that it is an ecological catastrophe with that particular plant. People are scarfing down tons of echinacea with no benefit. It is silly. (interview, 1999)

Although there is significant misuse of some less-dangerous herbs such as echinacea and goldenseal (usually in terms of people taking a lower-than-effective dose), no herbalist wants to see that misuse extend to some of the more powerful and potentially dangerous herbs (e.g., aconite, belladonna). These herbs are excellent tools in the hands of a highly skilled and experienced practitioner, but deadly cocktails to herbal "dabblers." K. P. Khalsa sees licensure as ultimately leading to a tiered system where herbs are graded and only appropriately skilled practitioners would be able to recommend those more powerful botanicals. He explains how licensure will solve this issue.

> I expect to see a tiered system develop. Nursing is a great example, where you have a nurses aide . . . then you have a licensed practical nurse, then you have an RN [registered nurse], then you have a nurse practitioner. People with different educational levels are given different responsibility. The nurse practitioners have almost the same responsibility [as] medical doctors. It is a good model to look at. . . . People will be able to understand what their limits are. Anybody should be able to brew up a cup of peppermint tea for their tummies. Probably anybody should be able to take echinacea for a cold if [he or she] learn[s] the appropriate use for it. If we could educate people and put it on the

> package how to do it and what it is used for, it will become culturally integrated. Then there are definitely certain things that should be restricted, an herb like belladonna, the deadly nightshade, which is used in doses of a drop of tincture. That is a very valuable professional tool, but I don't want the average person out there having access to it. (interview, 1999)

Finally, proponents of licensure for herbalists mention that professionalization legitimizes the practice to the mainstream, which, in turn, brings a higher standard of living for the practitioner. David Milbradt does not feel that licensure equals competence, but he does suggest that it would provide a higher standard of living for the practitioner.

> I don't think licensure brings competence as is commonly thought. What it does is allow a group of people to professionalize. Then maybe they can raise their standard of living as a profession. Then they are monopolizing the area they are in. They are restricting access for other people. That gives them a space where they can create a financial base. There is a real positive aspect to that. People should be paid for doing what they do. (interview, 1999)

Indeed, it seems that if herbal practice was formalized in some way, insurance companies might cover visits to the herbalist. The 2002 report from the White House Commission on CAM suggests that these two issues are closely tied together.

> States should, as appropriate, implement provisions for licensure, registration, and exemption consistent with the practitioners' education, training, and scope of practice. . . . A national coding system for CAM and periodic reports from federal agencies on coverage and reimbursement policy would further support inclusion of CAM in benefit programs. (cited in Bradford, 2002, p. 11)

K. P. Khalsa elaborates this idea, citing the need for third-party reimbursement in order to bring herbalism to a greater number of people.

As a person who has been in clinical practice for twenty-five years, I want to be able to participate in the banquet of health care services that Americans are using. I want people to be able to see me legitimately, aboveboard, to be able to communicate with [their] other practitioners, and have people be able to afford that service which means third-party payment eventually. That is a real sticky point for people. Their insurance is covered by their employer. There is a certain amount of out-of-pocket expense that they will tolerate, but there is a limit. We need third-party payment. (interview, 1999)

THE RESERVATIONS

Some herbalists feel that discussion of third-party payment and professionalization sounds too much like the biomedical practice of mainstream physicians. Herbalists such as Peeka Trenkle do not even consider themselves herbal doctors, but more "facilitators" for people who want to educate and heal themselves. One of the most significant aspects of holistic herbalism is the emphasis on self-care and grassroots independence. For these holistic herbalists, licensure or certification means more of a "doctorlike" role for the herbalist, which translates to greater dependence of the client on the practitioner. Peeka discussed her experience with the self-care system.

[Licensure] is a very bad idea because herbalism is . . . a self-care system. There shouldn't be licensure for people who are facilitators for self-care. . . . Myself, I am not licensed. I am completely classified as a layperson. But I have a very thriving clinical practice. I even hesitate to call it a clinical practice anymore. It is a consulting practice. People come in to learn. I have people coming in for all kinds of things. They learn, and they take that information and apply it in their lives and, lo and behold, lots of things happen—good things. But it is a system where it is the individual who is doing the work. It is a very different system. (interview, 1999)

Peeka discussed her belief in a vital force within each person, a belief that guides many herbalists. She feels that licensure by any established agency would be ludicrous as they do not understand the concept of vital force.

> The fact is that there is no pathway to licensure in this country that has any understanding of vital health. So, how can you expect a system that is so corrupt that at its core are the health and human resource department and the CDC [Centers for Disease Control and Prevention] and the FDA [Food and Drug Administration], people who have no understanding of even physical health, to set up a licensing system.... I mean, for heaven's sake ... there is no gestalt there. There is no understanding. There is no way that an herbalist would be understood. The kind of documentation, the kind of credentials and such that an herbalist would need to have would take them away from their work. There are so many people who, on the other hand, just because of how they live and what they know, are able to help people enormously. (interview, 1999)

Peeka's statement describes the essential incompatibility that many herbalists feel herbalism has with mainstream health-governing institutions. Herbalism greatly values its sense of spiritual connectedness and its use of intuition in healing. By intuition, I mean a relationship with plants and people that defies rational explanation. Matthew Wood (1997), another holistic herbalist, writes a poetic description of the irrational, nonlinear aspect of holistic herbalism and shows how conformity to the mainstream drive for rational explanation and, as an extension, licensure will possibly hurt the practice of herbalism.

> [The plants'] story would begin in the hidden depths of the Earth, in the unknown light of a place called the Underworld. From this source of mystery and power, the healing power and wisdom of the herbs ray out like a light, into our world. They never lose the magic that is innate to the inhabitants of a world of mystery, and in order to understand them as they are, we must adapt ourselves to their happy, mysterious selves. People try to press herbs into a rational, scientific box. This is not only entirely foreign to them, it kills their spirit and does not make ours soar. (p. 4)

Holistic herbalism has opened itself up to and begun to explore a great, new world of mystique that the mainstream does not even know exists for the most part. Although this is not the whole of herbalism, it is a key aspect, one that sets it far apart from the biomedical establishment and gives it the ability to open the mind of the American public and then dramatically shift the consciousness of those who are willing. Peeka Trenkle and many others feel that this value of intuition and other ways of knowing will be lost or compromised if professionalization becomes the focus within American herbalism. Orest Pelechaty, an acupuncturist and herbalist states, "You can't license for the intuitive aspects of herbalism. It is impossible. You can't even do very good in terms of educating on a standard level. It doesn't work well" (interview, 1999).

What you can license for is an understanding of anatomy and physiology, plant botany, biochemistry, etc.—all Western scientific ways of knowing. As mentioned in Chapter 3, an understanding of medical science has an advantage in that it helps herbalists communicate with allopathic physicians and may open up herbal healing to some who would not trust in an intuitive connection to the natural world. The AHG says, "It . . . seems imperative that herbalists seeking to practice within the framework of CAM practices must have some knowledge of conventional medicine" (Romm, 2002, p. 1). A fear on the part of those reticent to support certification, however, is that an understanding of biomedical science will become *the* benchmark for a successful, competent herbalist. If modern science is the agent of codification for the new herbal canon, overshadowing the vital force and intuitive connectedness, the power of herbalism, its unique potential to shift the dominant paradigm, will be lost. The potential of holistic herbalism bringing the sense of relational existence from quantum physics to modern biomedicine will be lost. If a system of professionalization is to be implemented, much concentration need be given to creative ways to circumvent the possibility of this loss.

Orest Pelechaty is hopeful, however. He feels that biomedical science might be changing to keep up with the innovations of quantum physics as mentioned in Chapter 3.

> I am a big optimist that what we have currently considered to be science or rather the perversion of science into reductionist materialism is not something that will stand forever. The scientific approach, which is to test hypotheses, has for a couple of cen-

turies been tremendously skewed toward left-brained, male-dominated, patriarchal, abstract, misogynistic, all that. It has been polluted by that. It doesn't intrinsically need to be that way. You can use logic and reason and experimentation to approach other phenomena that have been excluded a priori from this skewed, stunted view of reality. Science can indeed be expanded to include larger paradigms, nonmaterialist paradigms. There is that possibility. There could be a renaissance within science. It would be a big step to get the monied interest, the political interest, the corruption, the stranglehold of that from the scientific process, but I think that it is at least potentially possible. We can use a scientific approach, but a more holistic science. There is a hope for that. (interview, 1999)

It will take time for such changes to be manifested in modern biomedical science, however. In the meantime, the question of professionalization looms heavy on the horizon. This single issue has the potential to split the herbal community into two pieces, overshadowing the creative diversity and power of herbalism with an argument over certification, exams, and scientific understanding. Certification, although useful to some, may ultimately serve to limit the far-ranging practice of herbalism through codification of certain approaches to plant healing. A fixed system of certification implicitly demands that every herbalist declare his or her allegiance to the standardized or nonstandardized side, and silence will be just as revealing. Richo Cech wishes for unity within the herbal community, saying, "There is no 'us' and 'them.' Basically, we are all creatures of the Earth and we are all sustained by the oxygen that the plants give off" (interview, 1999). Such a nonpolarized attitude characterizes the idealism of the herbal community at present. The question of professionalization will polarize herbalists, however. When I asked Kate Gilday for her wish for herbalism she said the following:

My dream is to find a way that the herb world can stay intact. That the people whose heart and soul are really in working with the plants and appreciating all that they share with us and give to us that we can stay in communion with one another. There are several places that we are beginning to see rifts. They are just small rips in the fabric, but my prayer is that we get beyond ego and beyond fear and having to have the last word, or whatever, and truly work in the name of the plants. That is my prayer. (interview, 1999)

From my perspective, certification will certainly create and/or widen the rifts that Kate mentions. The proposed exam for certified clinical herbalists is designed to test competency in Western botanical medicine only (Tierra, 2002). In this way, herbalists from other approaches within the herbal community are immediately restricted from gaining this level of recognition. The AHG states clearly that it will not compel herbalists to certify, but if the public recognizes the certified clinical herbalist as the "competent" practitioner, then those who are not certified may appear to be less skilled regardless of the reason.

Part of the uniqueness of herbalism is the diversity of voices and approaches to healing from those who talk modern science to the Eclectic practitioners, from Wise Women to hybrid Chinese-Ayurvedic herbalists. That all of these individuals can sit together and evaluate a case study bringing slightly different aspects of a condition to light and pair that with a wide materia medica is enriching and enlivening. Each year at the Green Nations Gathering, a panel discussion of a case study is held. In 1999, Peeka Trenkle brought one of her clients who sat on the stage and fielded questions from the variety of practitioners beside her. The practitioners included Christopher Hobbs (TCM), Rosita Arvigo (Mayan healing), Kate Gilday (flower essences), and 7song (Eclectic, TCM, Ayurveda). Each used different diagnostic tools, came up with slightly different diagnoses, and built upon one another's treatment protocol. If certification by a standard method of practice is adopted, this panel discussion would not be nearly as rich as more practitioners will gravitate toward Western botanical medicine, the only approach that can be professionally approved. In fact, it would be quite boring to hear four practitioners say basically the same thing, in the same language, to an audience that is not learning anything new or earth-shattering.

Certification at its core is not about bringing herbalists' practices up to snuff or about bringing competence to an incompetent field. Most herbalists are already competent in their approach whether they use or understand the language of science or not. In many ways, certification is about centralization and control. It is about bringing the uniquely diverse practice of American herbalism under the yoke of biomedicine. It is about linear thought and distinctions—the Cartesian way of life. Herbalism represents and continues to develop a new circular, holistic way of thinking. Decentralization is, at its core, a holistic, circular quality.

If, as a diverse herbal community, we still desire a system of professionalization, perhaps more thought should be given to innovating a progressive, more inclusive structure. Is there a way to ensure safe practice without minimizing the diversity of practices and the role of intuition and mystery within American herbalism? Can licensure work from a decentralized perspective rather than a centralized review board?

I support the public safety concern offered in the section about licensure at the beginning of this chapter. At the same time, I feel American herbalism has much to offer that cannot be completely understood in terms of the dominant power structures and the methods of evaluation which we find in our culture at present (e.g., exams, essays, written applications). I also feel that to codify one approach to herbal healing is exclusive and implicitly says to the public that other approaches are not as valid. I realize that this is not the intent of the AHG, but I feel it is an unavoidable consequence of the push toward professionalization. American herbalists are more innovative than that; they are more inclusive than that. It would be interesting to see if there is a way to monitor the practices of clinical herbalists regardless of their approach. A standardized exam would be impossible given the diversity of approaches within American herbalism, but perhaps it would be possible to set up boards of certification that respond to either a particular region of the country or a particular approach within botanical medicine and holistic herbalism. This board could monitor the practitioners in the region and send reports to the AHG or some such organization. With fewer practitioners under the governance of each "decentralized" board, the certification or licensure process would become a more personal affair. Individualized attention is the backbone of American herbalism. Why should it not be the backbone of a certification process for herbalists?

With an individualized approach to certification, perhaps the applicant could select how he or she presents his or her understanding of herbal healing from a list of possibilities. Rather than a lengthy written application submitted to a central board, an applicant could have a monitor sit in on several consulting sessions with the practitioner, a site visit of sorts, followed by a personal interview. Perhaps, it would be a written report detailing an innovative healing technique that the applicant has pioneered, or a transcript of a talk given at a conference. Through decentralization of the licensure process, each applicant could be given personal attention and would be allowed to express excellence in the way that is most beneficial to him or her. This approach would be costly, but perhaps elected herbalists could volunteer their time to act as certifiers.

Regardless of how this is done, it is clear that professionalization is a divisive topic within American herbalism. As the growth in popularity continues for herbalism, now is a most essential time for solidarity. Now is the time for creative solutions rather than ongoing arguments.

Chapter 5

The Ecology of Herbalism: The Role of the Herbal Community in the Endangered Plant Crisis

> It is every American's birthright to be able to go into the wild prairies and see medicine plants growing.... (Richo Cech, interview, 1999)

Although it may be our birthright to visit wild prairies and observe stands of echinacea that have been established for generations, it is unlikely. What is left of original wild prairies is but a scattering of a once unending scape of grasses and native plants. The Jayhawk Audubon Society in Kansas (Neas, 2003) states that 1 to 3 percent of tall-grass prairie and 6 to 7 percent of midgrass prairie are left intact. In some midwestern counties, lasting sections of prairies are blocked off with signs that alert us to its beauty and preciousness. As for the echinacea, pleurisy root, and a host of other prairie plants, gaping holes in the soil from poaching tell us where they once flourished.

The example of the disappearing prairie aptly demonstrates how, as a culture, Americans have alienated themselves from nature, its bounty, and its beauty. Prairies have been replaced with strip malls and mountains with high-class ski resorts. We act for our own immediate pleasure. Although the satisfaction of pleasure is not necessarily negative, it becomes so when we act as if in isolation, when we deny our connection to the natural world around us. In *The Lost Language of Plants* (2002), Stephen Buhner describes the impact of this alienation from nature on humans.

> Once the Universe becomes a machine, no longer alive, once human beings are defined as the only intelligent life-form, a unique kind of isolation enters human lives, a kind of loneliness

> that is unprecedented in the history of human habitation of Earth. It is a source of many emotional pathologies people struggle with. . . . They internally denigrate and deny their most basic experiences of the livingness of the world in which they live, their connection to it, and the importance of that connection. (p. 51)

When humans lose this contact, this "bonding" as Buhner calls it, our attitudes toward the natural world change. We are no longer attached to nature. Nature is no longer an integral part of our daily existence.

It is ironic that our culture has begun a turn back to nature at the last possible minute: millions of square miles of abundant rain forest and prairie have been decimated; volumes of indigenous wisdom contained within tribal groups have been lost or exploited; and tribal people have either been murdered for their land or are on the verge of starvation because of an imported American diet (Arvigo, 1994). As we feel it all slip away, we begin to grasp for it, trying to recapture the beauty of nature and reestablish nature-based living in our lives. Ellen Evert Hopman shared her thoughts on this pattern.

> There is a complete lack of understanding and respect for indigenous wisdom, although I think slowly people are starting to realize this. It is because we are losing the rain forest; we are losing the indigenous people. [This loss] is starting to seep into people's awareness. Both as a pagan and as an herbalist, I have a tremendous respect for the ancestors, and everything that they have handed down to us. I am very interested in keeping that stuff alive. (interview, 1999)

People have begun to realize the role that nature could play in their lives if they open up to it. The National Center for Complementary and Alternative Medicine, part of the National Institutes of Health, has reported extensive survey results that show 36 percent of Americans are integrating some form of CAM treatment into their health care (Barnes et al., 2004). However, natural medicine does not necessarily make us closer to nature. We want to be "earthy" and connected, but we are still operating within the framework of isolation and objectification—the Cartesian, biomedical worldview. Thomas Berry (1988) describes human exploration of Earth and nature as autistic. We desire a relationship, but we have not yet recovered the language to develop true communication. He writes, "Our scientific inquires into the natural world have produced a certain atrophy in our human responses. Even when we recognize our intimacy . . . we cannot speak to those forms. . . . We have forgotten the language needed for such communication" (p. 16).

This skewed approach is perhaps more dangerous to the pursuit of nature through natural medicine than others because the object of our "autistic" affection is also the basis of life on this planet. That is to say, we are pursuing in our awkward way that which sustains us regardless of whether we understand the connection. We are fed by Earth and breathe the air that is produced by plants of the Earth. On a basic level, we are supported by Earth. Now we are coming directly at her with our large clumsy hands. This heartfelt, positive desire for connection, complicated by our culture's paradigm of dominance and objectivity, creates the present critical situation around endangered medicinal plants.

THE ROLE OF INDUSTRY IN THE ENDANGERED PLANT CRISIS

The American herbal renaissance is occurring primarily and most tangibly in the marketplace—the overflowing aisles of health food stores and co-ops and now even the corner drugstores and mainstream grocery stores. Nearly every store that has a health and beauty aides (HABA) section carries the top ten selling herbs—ginseng, echinacea, goldenseal, black cohosh, to name a few. These herbs, which are hyped in the media and processed into pills and capsules for purchase by the millions of Americans desiring natural healing, are most threatened. In this scenario of off-the-charts market activity, what is the role of the industry with respect to endangered plants?

Michael McGuffin, president of the American Herbal Products Association and former partner in McZand Herbals, presented a strict "business is business" logic in our interview. McZand decided to convert to agricultural production of certain herbs rather than wildcrafting when the company recognized that some of the hottest selling medicinals were declining in the wild; supply therefore could not be guaranteed.

> I have contracts for thirty-five acres of goldenseal. It was a strict, simple, business decision. A strict, simple, supply-and-demand decision. What I saw was my business growing potentially out of my control. As a businessperson, that's unacceptable. The only way to run a successful business is to control as many elements as possible. (Of course, everything is never in control.) I started seeing a potential solution long before I started seeing a problem. Clearly, with regard to supply; that is the place of the most significant concern. I wasn't that worried about the plants from which leaves are harvested. I'm still not. I was more concerned with those that are harvested for their roots and bark. I wasn't all that worried about sassafras, however, because even though you take the bark and the root, it's a weed. I was worried about goldenseal. I was worried about slippery elm. I was worried about echinacea. (interview, 1998)

McGuffin's emphasis was on strict, simple business decisions. Although involved with herbs, an alternative pursuit by trade, he based his decisions on conventional business practices. For McZand and other similar companies to compete on the national level, they must work within the mainstream business model based on principles of supply, demand, and the bottom line. McGuffin found a way to make those values fit an environmentally friendly philosophy, but his first concern was business.

When asked what responsibility McGuffin felt the herbal product industry had toward developing an ecological awareness of herbs beyond medicines, an awareness of herbs as living, growing, life-sustaining entities, he clearly answered that they had none. He responded, "I think the primary function [of the herbal products industry], especially in the light of the consumer self-medication, is to provide high-quality products and, as much as possible, to also provide meaningful, useful, not misleading information" (interview, 1998).

He added that as for the situation with endangered plants, the attitude of herbal product companies varied throughout the industry.

> Your question is, "What can the industry do?" Well, it varies on different products; it varies on different companies. What we did is we stopped using slippery elm and converted the slippery elm in all of our formulas to marshmallow root because we de-

termined that was a legitimate replacement. I don't think that's the appropriate response for a company whose entire business is identified with slippery elm. Nor do I think the appropriate response is for that company to hold a press conference to announce that they are going to go out of business, that they are going to stop selling slippery elm because it may be in decline. I think the only appropriate response from such a company would be to plant some slippery elm or start reformulating. I think that it is important to recognize that there are so many different perspectives. The simplest way to state the business perspective is that the first response to a supply-demand imbalance must be an increase in supply, not a decrease in demand. I am not interested, from a strictly business point of view, in running and telling people, "Would you please stop buying herbal products because maybe there's not enough herbs. We'll get back to you when we think there are." I think that would be bad for my company, bad for my employees, bad for the industry. I don't think that's a good approach.

So what can industry do? It gets to be very specific. I can only tell you what my company did. My company invested something like three hundred thousand dollars in the last four or five years planting *Echinacea angustifolia,* because that is the species we want to use, and goldenseal. We lost on echinacea. I know what it feels like to plow dollar bills into North Dakota dirt. It's kind of fun, but not for long. But we kept trying, and actually we did get thirty-four hundred pounds dried echinacea root through two different farms that we contracted three years ago. That came in the day before yesterday. (McGuffin, interview, 1998)

Ironically, although McGuffin's "business is business" attitude does not awaken a deep ecological consciousness, it does help the endangered plant crisis. The simple supply-and-demand issue that he took control of relatively early has caused McZand and several other herbal product companies to make commitments to grow their own raw materials, purchase only cultivated herbs, or put money into agricultural research. American Ginseng was potentially saved from extinction by businesses such as McGuffin's that decided to grow their needed material rather than continue to pay the skyrocketing market prices for wild-harvested plant material. Kate Gilday noted the same concerning black cohosh:

> Let's look at black cohosh for a moment, because if we hadn't started replanting black cohosh, my guess is that it would be gone in the next five to ten years. There would not be any more black cohosh. It is being dug out of the wild by the thousand pounds, maybe ten thousand pounds at a time. That is a lot of root. There is a great deal of woodlands in the East and the Southeast, but that is a massive amount of plant material. Now many larger herb companies are growing their own. They are doing all kinds of investigations and experiments in which they are growing the plant in the woods and they are also growing it in the fields. Researchers are watching to see how the growth works and how it changes. (interview, 1999)

Herbal product companies, especially those run by trained herbalists or plant people, are more environmentally conscious than they were several years ago, but there is still concern over larger companies that utilize herbs in their production. Richo Cech relayed the observations of a plant conservation researcher:

> There is, however, another faction of wildcrafters who are working for big companies. We were just getting some information yesterday from a researcher who said that she had watched the trucks [from a major soft drink manufacturer] unloading *Chimaphila umbellata* (pipsissewa) by the bale. We all know that pipsissewa is something which, like most wild plants, is locally common or locally available, but when you go to the fringes of its habitat, it is much more susceptible to any picking. *Chimaphila* is susceptible to picking even if you are in the midst of a big stand. Two whorls have been picked off the top of a plant, which usually has three or four whorls, and we have seen that after four years there is only half an inch of new growth. It is extremely slow-growing. Even if someone is out there respectfully taking a quarter of the patch, it is a scar, which is going to be visible for a decade or more. When a big company goes in to pick pipsissewa by the truckload, you know they can't be giving it that kind of care. They are going to be pulling the plants up by the roots, swishing them off in a tub that is next to a truck some-

where, laying them out to dry, and then baling them up and taking them out for extraction for flavoring purposes. (interview, 1999)

Such abundant disregard for plants is not unusual in the herbal products industry itself. Growth and consumer demand compel companies to harvest larger volumes of plant material. With these expectations, businesses are not necessarily cognizant or caring about the environment even when the natural environment is the source of their wealth. This is compounded by the fact that larger companies with larger supplies of capital are generally able to cover more ground searching for plants and are able to "gather" a higher yield by utilizing mechanized strip mine tactics. Herbalist and conservationist Deb Soule describes what she sees as the consciousness that allows these practices.

At this point, we are in a time where the bottom line for so many people is money. The relationship to the natural world and to the health of the ecosystem as a whole is not at the top of the list for most. (interview, 1999)

In regard to harvesting herbs for large-scale production, a note must be made about individual wildcrafters who gather medicinals using the traditional trowel and clippers method instead of mechanized harvesting. The herbal community's outlook on this group fluctuates, but at this time many herbalists regard the wildcrafter as practicing sustainable harvesting techniques, and therefore individual wildcrafters are less of a factor in the endangered plant issue. Richo Cech has stated:

I don't think that it is the responsibility of the wildcrafter or it is the karma of the wildcrafter population, that we are running out of wild plants. I think that there is a certain dent that collection for medicine puts in wild populations, and that this can be done sustainably on a bioregional level with people making medicine for their family and friends in need. I think there is a bigger dent put in it by certain larger companies and corporations coming in and harvesting on a large scale. (interview, 1999)

Kate Gilday pondered the issue of wildcrafters bringing it back home.

> There are people in the Southeast that I know of who are gatherers of the wild plants and have been for generations. These are people in Appalachia, people in the Smoky Mountains. These are people who harvest for their livelihood. It gets tricky at that point. (interview, 1999)

In 1995, United Plant Savers (UpS), a national grassroots medicinal plant conservation organization, was formed by a group of concerned herbalists, many of whom participated in this study. In outlining the values of the organization, board members decided to stand firm on a refusal to issue a moratorium on the sale and purchase of wild-harvested endangered medicinals (UpS, 1999). The reasons they provide are that to take such a strong stand would (1) require UpS to dedicate time and money to policing the moritorium, (2) endanger the livelihood of bioregional wildcrafters, and (3) alienate UpS from many business entities who did not share the organization's concern. The companies might be forced to make changes, but UpS would prefer that they do it because the company understands it is a good thing to do. In this way, UpS hopes to be more of a consciousness-raising organization than a policing entity.

This is certainly a noble position to take, but it falls short of the mark. Although it would certainly take resources to "police" a moratorium, issuing such a statement would heighten awareness around the endangered-plant crisis. Even *suggesting* that a moratorium be issued would have an effect. In my own practice, I decided to eliminate use of endangered botanicals until I could find a reliable organically grown source. The same has happened in many other herbal-product businesses. This practice does not translate to all herbal business however. As Michael McGuffin suggested, businesses are not obligated to and are unlikely to make decisions based on ecological concerns if their bottom line is not imminently threatened. What UpS relies on in not calling for a moratorium is a reshaping of the profit-and-loss business paradigm. It might be more advantageous to issue a moratorium for the present that could be lifted at a later date. At that point, conservation organizations could work on changing the mindset and motivation of the herbal products industry, for deeper, long-lasting healing.

As mentioned, many herbal product companies and a host of smaller, regionally based companies understood immediately (and perhaps even before the advent of UpS) that they had a potential problem with the supply of wild-harvested medicinals. Many of these companies are able to rationalize this from a business standpoint as well as from an ecological mission and vision standpoint. Many are vertically oriented, meaning they are out in the fields and woodlands searching for and growing the plants themselves. They are in touch with the plight of endangered medicinals and are doing their part to remedy the situation by converting to organic cultivation methods, changing formulas, or discontinuing products altogether. Although industry on this level continues to develop its ecological ethics, such companies are not the ones who need to be scrutinized. More dangerous are those companies that manufacture widely marketed brands that end up in the mainstream supermarkets and drugstores, and other companies that use natural, herbally derived flavorings in their products. Such businesses often use impersonal harvesting techniques coupled with a short-term environmental outlook.

UpS can wield a great deal of influence over mid-size and smaller companies. The heads of such companies are herbalists for the most part, and are networked in the same circles as the board members of UpS. They respond to the bottom line, as Michael McGuffin said, but they also respond to environmental consciousness much more readily than do some of the larger, multinational corporations. For the larger companies, however, the profits are astronomical. They are not herbalists. For these companies who operate out of a different paradigm than the herbal community and its herbal product companies, a moratorium and boycott might be the only language that they understand. In this way, United Plant Savers' resistance to issuing a moratorium will hinder the ability to conserve threatened medicinal plants.

WHAT ABOUT BEYOND INDUSTRY?

It would be easy to say that the herbal product industry is the sole player in the endangered plant crisis, but if we are to be truly holistic about the situation, if we are to apply the values and ideals of herbalism consistently to the health of the ecosystem, then we need a more precise understanding of the paradigm at work. It is not productive to attempt to lay blame. As was noted in Chapter 3, cause-and-effect, subject-object relationships do not accurately describe what is happening. Rather, we need to view this critical issue as the culmination of a dynamic system of phenomena. I proposed at the outset of this chapter that there exists a deep cultural coding that conditions American interaction with nature. With that background in mind, let us examine two specific aspects or subissues within the endangered plant arena: monopopularizing and the "domino effect."

Monopopularizing

Monopopularizing refers to the consumption of one or a few medicinals out of the much larger and comprehensive range of the North American materia medica. Herbalists have at their disposal a multitude of plants for use in healing, all with subtly different indications, but only a select few of these make it into the media and then onto the shelves of mainstream stores. St. John's wort, echinacea, saw palmetto, ginkgo, ginseng, and black cohosh are by far the most popular (i.e., most money producing) herbs in America today (Blumenthal, 2003). In the media they are touted as amazing, quick cure-alls with "magic bullet" properties. The primary function of this phenomenon is to give the mainstream a grasp of herbal healing. Richo Cech spoke of this situation.

> I think that the idea of monopopularizing comes out of a "this-for-that" kind of consciousness, which is really prevalent in the modern context. Generally speaking, people are taking a complicated study—herbalism—which relates to the use of potentially thousands of species of plants worldwide and putting it in popularized terms. The "top ten" plants is a common way to work with that. (interview, 1999)

Monopopularizing directly threatens the health of the ecosystem by putting enormous stress on just a few plants instead of distributing the desire for healing agents over a wide variety of them. Some of the more popular medicinals are harvested for their roots (e.g., ginseng, goldenseal, echinacea, black cohosh), which grow slowly and are harvested in nonrenewable ways (i.e., without replanting the crown of the root).

To begin to remedy this issue, we must understand the systems at work here. First and foremost, we have a public that is at the beginning stage of learning about medicinal herbs. As Richo Cech mentioned, herbalism is a complicated study. If one is not familiar with it, it can be overwhelming. In that sense, it is understandable that the mainstream public would select only a few herbs to begin their study. As an herbalist, my first year of study involved investigation into ten plants. I grew, gathered, mediated with, and read about those plants in depth. At the end of that first year, I had established a level of trust and familiarity with those plants. I anticipate that as Americans become more comfortable with herbalism or botanical medicine, they will expand their materia medica.

The second factor in this issue of monopopularizing is the mainstream desire for a magic bullet. A magic bullet outlook is residue from the use of pharmaceuticals: one drug does it all immediately. In herbalism, a health issue will often be approached with a formula of herbs, each of which has a slightly different action and addresses a layer of the condition. Furthermore, although some herbs have dramatic, immediate effects, it is more likely that a formula would be used for an extended length of time and would have a gradual and cumulative effect. In searching for an herbal magic bullet, the mainstream public has reached out for some of the stronger, more dramatic herbs (e.g., goldenseal). The public wants the strongest herb with the broadest application. Therefore, they have narrowed their materia medica to only a few plants. James Green summed this up.

> Monopopularizing is kind of like the cousin of magic-bullet consciousness, right? They are doing the same thing—here, take this herb and it'll do this. That's all: that reductionist, magic bullet, treat the symptom with this thing. That is what is being marketed; that is where the marketplace is. (interview, 1999)

The third element that creates monopopularizing is the hype put out by the media. The media seeks to feed and fuel the mainstream curiosity of herbs. They do this through front-page articles in *Time, Newsweek,* and various national newspapers. Such articles generally discuss the top-selling herbs in druglike terms. The November 23, 1998, cover of *Time* pictured echinacea, ginkgo, St. John's wort, and ginseng growing out of a prescription bottle. Such coverage serves to increase the public profile of these few herbs (as well as the magic bullet attitude) and neglects the many other more prolific, abundant herbs. The media encourages the mainstream to buy more herbal products from the limited top ten materia medica. Michael McGuffin touched on this in our discussion.

> With regard to monopopularizing, the herbal products industry doesn't do that. The media does that and then we try to keep up. We don't go out there and say, "The only thing you ever needed for anything having to do with the female reproductive system is black cohosh. Please run it on the seven and eleven o'clock news." The media picks up a little story and they run with it, and yes, then we try to keep up. Then we do try to meet those sales. We didn't decide not to. We didn't look at it and say, "This is a problem. This is a small plant. We don't know what the supplies are." We simply haven't caught the wave yet. I don't know if there are any plans in the marketing department to do that. I know that if I am aware of that I would certainly push all of the soy research on them that I could. But I know what they would say: "Yeah, but *USA Today* ain't selling soy. *USA Today* is selling black cohosh." That's a tough thing to ask a company. (interview, 1998)

Nascent mainstream interest, the search for the magic bullet, as well as media hype and control contribute to and form the potentially devastating practice of monopopularizing. How can we constructively work to remedy this complex? Rosemary Gladstar gave some advice to newcomers in the field of herbal healing—advice not often heard in the mainstream media.

> I think that the very first thing people should know about herbs is that herbs, like everything else on this planet, are not an endless resource. If you are even going to tinker in herbal medicine, even start to investigate it, the number-one issue is understanding the plants in relationship to their communities and native environments. People don't want to take that approach because they say, "Ahhh, we have to learn about dosaging and what's safe and contraindications. Now I have to be environmentally responsible also?" My basic message to people is that if you are not willing to be environmentally responsible, then use drug medicine. We need to have an educated public in terms of the en-

vironmental issues surrounding herbalism right now. That is my personal agenda. (interview, 1999)

Thankfully, more alternative media and herbal trade publications are giving attention to the environmental aspect of herbalism as Rosemary suggests. Perhaps writers for mainstream publications could pick up this trend and spread the word to more casual users of herbal medicine.

Beyond environmental education, the mainstream needs to understand that herbalism is not a consumer-driven practice, although the herbal renaissance at present is. In fact, it is quite the opposite. The beginnings of the American herbal renaissance (with the back-to-the-land movement) came in part as a reaction against widespread consumerism and market dependency. For many herbalists, herbalism is most directly about independence from the marketplace. Peeka Trenkle commented on this.

> [Y]ou can't have freedom if you don't take responsibility. That is a basic premise. You don't get the responsibility or the freedom by simply going to a health food store and buying your herbs. You can do this if you already understand freedom and responsibility though. I can go to the store and buy ginseng if I need it. It is nice that it is conveniently there, but I can also do without any products. I can also go in my backyard and make myself a tea from what is growing there. My vision of herbalism is that we have more herbal endeavors, group endeavors and have people have their own herbs in their backyard. Those people who live in the cities can all share a community spot and grow their own plants and make their own medicine. Or they have their neighbors make their medicines. We start to manifest more community with each other and work bioregionally to make that part of our lifestyle again, to make the Earth part of our lifestyle again. (interview, 1999)

If herbalism is utilized in its full ideal paradigm, the subissue of monopopularizing ostensibly disappears. Although herbalists use many of the same plants across the country, many of these are abundant, weedlike plants. James Green says, "It might be okay to monopopularize dandelion because there is a ton of it out there. Perhaps if this pattern of marketing is to continue, herbal product companies should concentrate on those herbs that are in abundance" (interview, 1999).

At the same time, the herbal community itself is becoming increasingly conscious of this issue and is working to educate people about the wide array of healing plants growing in backyards and along roadsides everywhere that may be harvested and made into medicines in the kitchen. Areas where people generally think "nothing" grows are home to many useful plants such as coltsfoot, mullein, yellow dock, St. John's wort, and horsetail to name a few. Ryan Drum emphasizes the connection between the public's awareness of these resources and the consumerism that drives the American herbal renaissance.

> The market phenomena . . . of merchandising and selling herbs capitalizes on a fundamental need which is antimarket, in that the herbs are available—many of them just for the taking in waste and wild places everywhere, the places that can't be farmed. The lack of knowledge about how to get them and harvest them at the right time, the right place, and use them leads to market dependency. (interview, 1999)

The Domino Effect

The other subissue within the endangered plant crisis that requires examination is the so-called domino effect. This term refers to the practice of substituting a nonendangered plant for one that is endangered because it has similar actions or active constituents. It is called the domino effect because as an herb becomes endangered, its popularity is shifted to the next, most similar herb, which is likely to become endangered as well. In this way, the result is akin to a line of dominoes that lean on one another and so collapse the whole line. As a plant becomes threatened or endangered, herbalists have taken to recommending one or several other plants that would achieve similar clinical results. Although this practice is certainly noble, it may produce undesired results. Substitution does not address the root cause of the problem, but rather attempts to divert or delay any trouble. The end result, the knocked over stack of domino herbs, however, is more deleterious than the initial issue.

Widening the range of the mainstream materia medica is essential to preventing or healing the endangered plant crisis, but substituting does not accomplish that aim. It simply replaces one herb with another herb. As will be addressed in more depth to follow, the substitution of herbs requires that one look at the herb as a set of chemical constituents viewed in isolation rather than a whole plant with a dynamic and synergistic healing effect. By widening the materia medica, we may increase the number of available plants and, therefore, the subtle qualities that may be used to address the equally subtle human system. At that point, the system of healing becomes much more precise and sensitive.

To understand this more fully, let us examine the domino effect around goldenseal. Goldenseal is often touted for its berberine content, which has strong antimicrobial action with an indication for congestion of the respiratory tract (Mills, 1991). For that reason, goldenseal finds its way into many cold and flu formulas, one of the most popular classes of herbal products (Brevoort, 1998). Goldenseal happens to be a slow-growing forest plant with a particular environmental niche. Since it is harvested for its root, the plant is usually killed as a result of harvest, although one may replant the crown of the root. It takes three to four years for a single root to come to harvestable maturity (Blumenthal in Gladstar and Hirsch, 2000). This set of factors—strong medicine, popular medicine, slow growth, particular niche—make goldenseal one of the most highly threatened native plants of North America.

When herbalists realized that there was a potential problem with goldenseal, they recommended that people utilize other plants that contain the same berberine alkaloids and therefore produce the same therapeutic effect. These plants include Oregon grape and gold thread. It happens that Oregon grape is another slow-growing medicinal and gold thread, with its particular forest niche, though locally abundant in some areas, is nationally scarce. After only a few years of this substitution policy, both Oregon grape and gold thread populations are in decline. Both are listed by UpS as potentially threatened.

What is happening here? Kate Gilday summed up the attitude that brought on the domino effect.

> The thought was at the time, you should use the berberine-rich plants growing where you live rather than using the goldenseal. In the Northwest, it was Oregon grape root. In the Northeast, it was gold thread. Down South, it was yellow root. Those were the native plants of those areas. People started using those plants

instead, and lo and behold we used too much of those too. The thing about these plants that you are looking at is that most of them have a very slow growing period. Some take a long time to flower. Oregon grape is a good example of a slow-growing plant. Gold thread needs its own particular climate and environment. They are all native plants, and they all need something specific. Yet we are harvesting them, and they are taking "too long to recover." (interview, 1999)

Deb Soule elaborates with a wise analysis of the problem.

> We can reduce a plant to just its chemical constituents, but perhaps what we need to be doing is looking at the bigger picture. What plants grow where we are and why is it that we use certain herbs and not others? For example, the slippery elm issue—people try to use marshmallow root instead of slippery elm. We are after demulcent properties or [in the case of goldenseal] the berberine properties. . . . I think that it is important to keep that in perspective, but I also think that we have to think a little bit bigger about it. The difficulty, and maybe that's again where the paradigm shift has to occur, is in helping to educate people that it is not about using St. John's wort in place of Prozac. It is not about using gold thread instead of goldenseal because of the berberine content. It has to be a bigger paradigm shift. I think we have to include more than just the constituents in our reasons why we choose certain herbs. I think it is important to understand that aspect, but I think we have to become a bit bigger. (interview, 1999)

In substituting one popular plant for another, the herbal community has done what it claims not to do or seeks to avoid—it has begun to observe a plant as an active constituent. Rather than attempt to directly defuse the issue, we have further complicated it by falling back into a Cartesian worldview—a plant is a carrier of a certain constituent that can be observed and isolated. We can simply take one berberine "carrier" rather than another—this for that.

One solution to the domino effect is presented in *Planting the Future* (Gladstar and Hirsch, 2000), a volume of essays on endangered plants written by holistic herbalists. In the chapter titled, "What You Can Do to Make a Difference," using "herbs for wellness, not just illness" is emphasized (p. 15). Many of the plants on the UpS at-risk and to-watch lists are strong medicinals used for particular disease conditions. The suggestion is made that rather than using herbs when one is ill, one should consider herbs as a way of life. This means, in part, using some of the more abundant, weedy herbs as wellness tonics in order to foster health. In this way, illness is reduced and, in turn, the amount of endangered plants used is also diminished.

Deb Soule and Kate Gilday would suggest that another solution to this issue is to concentrate on plants within your own bioregion. There is a living pharmacy close by every person in this country if he or she desires to learn about it. It would be interesting to analyze sales data by bioregion and, using an estimate of the plant material in each region, calculate if bioregional herbalism could safely satisfy consumer demand. My assumption is that it would, but I think that is further down the "paradigm shift" line. It begins to venture into ideas of decentralization, which demand a book or several onto themselves. As for bioregionalism at present, Annie McCleary spoke briefly about osha, another endangered herb, and a way to approach its national scarcity.

> My feeling is that habitat everywhere is being compromised by pollution and agricultural practices and development. We all know that. There is a lot of osha out there; some of it is on very protected land, *but there may not be enough osha for everyone everywhere.* (emphasis mine; interview, 1999)

Bioregionalism is a double solution. Not only would it begin to take pressure off of nationally scarce or slow-growing plants, but it would also begin to extract plant healing from national consumerism, a situation that Ryan Drum spoke of earlier in this section.

Although discussion of the domino effect as it pertains to goldenseal shows how easily the herbal community can slip into short-sighted, Cartesian thinking, I must include praise for the development of whole-plant mentality in other areas. In the case of St. John's wort, the herbal community has almost universally asked the mainstream to look beyond the noted constituents, hypericin and hyperforin, and examine the plant as an entire entity. Furthermore, herbalists often suggest other plants (oats, lemon balm, motherwort) that may be more appropriate holistic medicines, for the treatment of nervous system issues. To allay the domino effect, the herbal community must consistently observe the value of the whole plant as opposed to the individual constituents.

ARE PLANTS REALLY ENDANGERED? AN HERBALIST'S PERSPECTIVE

I sought out Kate Gilday for an interview on endangered plants because to me she symbolized the Earth mother. Her voice and mannerisms are soft and subtle, her laughter rich; she obviously has a deep relationship with Earth that brings a calm joy to her every thought and action. She is an avid gardener, wildcrafter, and woods woman. I felt that she would have amazing insight into the endangered plant situation and would be able to make the necessary links to our attitude as humans toward Earth. She brought all of that to the interview, but first she threw me way off and started my mind churching. She began with the following.

> One question in a series that has come up for me is, Are certain things really endangered or are we not going out to where they are? Are they still around but in quieter communities that are deeper in the woods? Does that mean we should leave them alone? Or does that mean we can harvest, in a good way, a small amount? (interview, 1999)

I gasped. Are plants really endangered? Of course they are. If they are not, what are we going to talk about for the next hour? I asked her to expand what she was saying.

> I don't know if it is so much that the herbs are retreating farther into the woods as much as we only go so far off the trail as humans. Most people—not just the general public, but even the herbalists—do not go very deep into the woodlands or go beyond a certain geographical area in their findings. (interview, 1999)

Could that be so? I understand that the herbal renaissance is a marketplace phenomenon, a primarily industrialized pattern, but could that extend into my ideal world of herbalism? I thought of the other people whom I had interviewed. Were they part of that complex? Did they stay on the "trail" literally and figuratively? That is a tough judgment to make. My sample population is wide and varied. About 50 percent harvest, process, and use their own medicines or get them from a closely related source. But that is only half of a population that I would assume would make these activities a greater part of their daily lives.

It seems that Kate is partially correct. Perhaps not only are native plants retreating to a "safe space" in the forest, but herbalists are also retreating into the industrialized, consumer medicine model. Perhaps we are not going deep enough into the woods, into the wild. Perhaps we are not going into the woods at all but rather picking up the phone and calling across the country to order dried plant materials.

I include this perspective to give another view. It challenged my own thoughts, and I hope that it will make the reader think critically about his or her place and pattern in Earth space. I have to say that following my discussion with Kate, I made it my mission to venture farther and farther off the trail in search of plants and nature experience. By journeying out, not surprisingly, I was able to interact with plants I thought I would never see in the wild.

Whether one feels that plants are retreating into the woods, leaving the planet altogether, or simply that the intense marketplace activity is threatening the numbers of certain plants, one thing is clear: it is time for humans to recover an interaction, a two-way relationship, with the plant world, to "bond" with Earth as Stephen Buhner (2002) says. It is certain that curiosity is present throughout ever-expanding pockets of the American population. Mainstream acceptance of and concern for nature is widening and is manifesting through the herbal renaissance and growing environmental concern. To ensure that the current at-risk plants do not disappear, that potentially threatened plants do not have to experience the scenario of extinction, and that the ecology returns to balance, humans need to open up to personal relationships with nature that they find around them. By that I do not necessarily mean that people have to spend their days in meditation directly communicating with plant beings. For some that is the most appropriate and comfortable avenue. However, everyone finds their own way of entering into relationship and allowing the plants to affect their being. Perhaps it is through gardening (on a windowsill or in a field); perhaps it is through walks in the local park where one gets to know the names of the plants along the path. Whatever allows one to feel the importance and significance of herbs not just as medicine, but as living breathing entities, is appropriate. Kate Gilday says that it might be that tending to and harvesting from a wild patch of at-risk plants is part of that relationship. She explained her interaction with blue cohosh in her area. She felt that to selectively harvest from the patch was the best solution.

> [I]n some areas near to us, we have a local abundance of blue cohosh. Do we harvest blue cohosh in this area? Yes, we do. It is healthy for the plant populations that are there to sustainably harvest so that they can continue their growth cycle, their abundance, their reproduction. In that way, I think it is almost foolish not to harvest. We tend to them like our own gardens. We are watching every year, watching the development of this bed. This one area that we harvest from is five acres of blue cohosh. That is a lot of blue cohosh. What we take isn't even a dent. We may take six to twelve plants and we replant the crown. What we are finding in some areas is that the blue cohosh begins to die out if it is not thinned because it is strangling itself. The same is true of goldenseal. If you have a mother plant, after a number of years you have all of these baby plants springing out from her, but her energy has been drained by all that reproduction, and then the plant itself dies from the center out. This is in contrast to the growth patterns of ginseng, which will grow for years and years and years as a single plant. These types of situations make the educational component so essential. By taking some of the smaller goldenseal or blue cohosh plants and some of these two- or three-year plants and replanting them elsewhere or using them in medicine, you are actually helping the health of the bed you are working in.

> You have to have that commitment to tend that area though. As a wildcrafter, I feel committed to that. That is what I need to do to feel good about harvesting this plant, any plant in the wild. Even if it is an abundant plant, I still feel it is my obligation, it is my duty, my commitment, my love, to help them to continue their cycle in the best way that I can. Sometimes it is by gathering, and sometimes it is by propagating. (interview, 1999)

If people cultivated the kind of relationship with plants (especially those which are endangered) that Kate has with blue cohosh, it would be more harmonious to harvest than to delineate certain plant populations as off-limits. Such forbiddance seems a bit artificial if people can be responsible and interactive. Deb Soule succinctly summarizes the point that we need to understand and respect plants as vibrant, full-of-life beings with purpose and reason just as humans. She says, "[Plants] are amazing living beings with as much right to be here as we do" (interview, 1999).

If we as a species and particularly as herbalists can understand this point fully, our environmental and ecological concerns will become part of our everyday working toward greater life sustainability. If herbalism can live its ideals, and spread an example to others who are just putting out their "true leaves," so to speak, discussion about endangered plants could be irrelevant. The herbal community has begun an ecological awakening within itself; with constant work, focus, and critical thought, a true two-way relationship with ecology can become a reality. From there the herbal community can serve as a teacher to the society at large.

Chapter 6

Sinking Roots, Reaching Branches: Spirituality and Tradition in Modern American Herbalism

>I think it may be similar to a lot of things that have happened in the last two hundred years and in this century in that our knowledge outstrips our skill level. We are able to make bicycles, and it took a long time before people learned how to ride and use them correctly. Then came the automobile. The first twenty years of popular automobile use were really risky adventures. Just going out for a two- or three-hour drive could end up with a two- or three-day set of blowouts, gasoline leaking, gasoline with water in it, all of these things. The understanding of what would make it work smoothly was not there. (Ryan Drum, interview, 1999)

Although the present herbal renaissance is not the first in American history (Griggs, 1997), it is the first to sketch out a comprehensive view of health with regard to the individual, the community, and the planet. This is certainly noble, but the contemporary rise in the popularity of herbalism has been marred by the bad gasoline, poor road conditions, and tire blowouts that Ryan Drum alludes to. The previous chapters detail three such "potholes" (the inclusion of science and folklore in herbal practice, the debate over licensure and professionalization, and the crisis surrounding endangered medicinals), places where the herbal community must develop a firm, united, critically examined long-term vision. The herbal community as a whole lacks a clear, unified understanding of what would "make it work smoothly," what would begin to give herbalism deep relevance to today's society. That understanding includes an integrative, deeply rooted spirituality that would be accessible to a wider audience.

Although pockets of the herbal community have significant spiritual practices or relationships in their work, this is far from universal, and as American herbalism becomes increasingly popular, this will be diluted even further if strong leadership and direction (from the AHG or a decentralized review board as well as from herbal teachers working with those new to the herbal community) are not taken at this moment. I do not write this to belittle or underestimate the deep connectedness that many American herbalists have developed with land and the spirituality it holds. Certainly many participants in this study already feel the desire and the need for connection to a spirit force (or forces) in their own practices and their interactions with the wider world, and they have manifested it. This work is meant to solidify what already exists and encourage that which is forming through an analysis of the community and a synthesis of its many "voices." It is my sole hope that American herbalism will be a relevant, sustainable part of the twenty-first-century view of health and healing, and that it will awaken the sleeping American mainstream consciousness and help it realize the important links between the individual, community, and ecology.

At the outset of this project, I had a long-distance discussion with ethnobotanist and shaman Enrique Salmon about spirituality in America. My premise was that Americans, as a culture, lacked a sense of spirituality and were destroying their environment in part because of this. His argument to me was that America, nonnative America, had a deep spirituality; it was simply not Earth-based. One could continue that argument, and perhaps it is without a clear resolution. However, as a born and raised Christian, my experience of spirituality was limited. I was exposed to a great deal of religion. I define religion as an established, coded system of thought that attempts to connect individuals with an external force and thereby give meaning to life. Although spirituality, based in a relational framework, connects the individual to his or her community and place in the cosmos, religion, as Matthew Fox (1983), Christian theologian and scholar, says, is based in a "fall/redemption" philosophy. Religion, he goes on to say, "is a dualistic model . . . fall/redemption [philosophy] does not teach believers about justice-making and social transformation, or about Eros, play, pleasure, and the God of delight. It fails to teach love of the earth or care for the cosmos" (p. 11). Fall/redemption connotes linear, cause-effect thinking. For example, Eve eats the fruit of the forbidden tree and is punished with exile from the Garden of Eden. In this way, the underlying logic of religion is quite similar to that of biomedicine. In contrast, the use of nonlinear, relationship-based philosophy in spirituality is akin to the essence of holistic herbalism. In this way, holistic herbalism is to biomedicine as spirituality is to religion.

America is not a spiritual country, but rather a country of diverse religious practices. Spirituality is quite the opposite of what American religions provide. From my point of view, spirituality empowers individuals and seeks to connect them to their surroundings in a meaningful way. How can one clear-cut a mountaintop if one understands that it is the home of God? How can one dominate others through material drive if one understands those people as images of a living God?

A potential source for this disconnection between religion and spirituality stems from the American "melting pot" syndrome. Excepting natives to this land, Americans are descendants of immigrants. When asked where my ancestors came from, I answer that they were Irish and Italian. We as a country have left the deep meaning of our stories and rituals as well as our connection to a "sense of place" in our ancestral homes—Ireland, Italy, Zaire, Haiti. . . . Computers connect us to farther and farther reaches of the globe, but they do not connect us to that deep knowing of land and community, that experience and sense of place. Kate Gilday elaborated on this.

> My partner, Don, and I have talked about this for hours at a time. People of North America, native people who have been here for thousands of years, this is their home. This is what they know. This is what their bones know. It is what their hearts and souls know. When they talk about ancestors, it is direct. It is right here on the ground. . . . It is in these woods. When you sit with people of native descent, they are totally at home in themselves, in the land, with each other, with all of nature that is around them; they just know this place. I feel like those of us from European descent, first of all, have lost our connection to home. We have lost our connection to ritual that came from wherever we were. I mean, every culture had ritual, every culture had tradition, every culture had ways of celebrating the seasons and the Earth. That

got ripped away when we moved from the old land to this land, North America. (interview, 1999)

As we have disconnected ourselves from our native lands or have been forced to move by oppressive, governmental powers, we have lost the place where we find home. With this disconnection, we, as immigrants, have lost the relationship with the land that is known in our bones, as Kate says. This disconnection is analogous to the situation many people experience when they move to a new place. In moving, one leaves one's community of friends and family. The process of establishing a fresh circle of friends in a new place takes an extended time. Developing the level of trust and comfort that one found in the former community is not automatic. When one lives with a community for an extended period of time, a common language is developed in which many things are understood implicitly. In the new community, that common language must be rebuilt from the ground up. In this way a certain sense of isolation and disconnectedness is common.

We have also, by extension, lost a sense of how to integrate the use of herbs into our healing paradigm. The "technology of independence" of which James Green speaks needs to be relearned. Every culture across the globe had at one time a system of healing involving native-plant medicines and therefore people had a context into which they could place the practice of herbalism. Systems of healing and cultural context are manifestations of a common language developed between people and the land. Just as in the example of a person moving to a new area, common language comes out of a long-term relationship between the two parties. Americans lack this sense of relationship, creating chaos and unfamiliarity around the herbal renaissance in this country. K. P. Khalsa outlined this issue.

> [Herbalism] is a systematic approach to the use of herbal medicine, which is honored in every culture, everywhere around the world. Every established culture has an herbal medicine that is many centuries old. . . . In our culture, we are a bunch of immigrants who came here and left the established paradigm of herbal medicine from our particular ethnic background . . . wherever it was.
>
> Now what we have in the new herbal gold rush is just a chaos of manufacturers dominating the public consciousness. . . . Peo-

> ple don't have a historical context for herbalism. Most people's experience of herbal medicine is pretty skimpy. They get a cold; they use some echinacea. Then they say things like, "Well, it worked great because instead of my cold lasting for ten days, it lasted for eight days." For an herbalist, that is just stupid. We want to see their cold gone in twenty-four hours or we know we didn't do the right thing. You don't just blast away at it with small doses of echinacea. We have to have a holistic context to put this in. . . . That is herbalism. (interview, 1999)

How do we make sense of this marketplace chaos? In order to remedy anything, we must understand the motivation of the people interested, we must understand what they seek. Again, Kate Gilday provides insight.

> We are seeing now the renaissance of people going back into native, indigenous cultures and trying to glean from them what it means to be connected. We are trying to figure it out. What is our land? Who are our ancestors? Where do we find "home"? I think we often find home in our own hearts, but at the same time I think that the land and the trees and the woods and all that is around us, the waterways, can really be our home if we can just open ourselves to those experiences. There is such resistance. And fear. In our classes, we try to help remove that fear and separateness so that people will recognize the connection to the woods and to the native plants, the connection to this place being their true home. (interview, 1999)

If this is indeed what Earth-seeking Americans are searching for, if they desire a true connection to their surroundings, the American herbal community has a great responsibility. Holistic herbalists, in particular, realize these connections and understand the importance of "place." The responsibility of the herbal community then is to help to refine these connections into a practical discipline with rich two-way relationships between participants (herb, herbalists, enthusiast, client, etc.). However, a culture forms and affects the healing paradigms within it. Indeed, the field of medical anthropology is constructed to further understand these connections (Landy, 1977). Herbalists form their own unique subculture, but they are still products of the overarching American culture. American herbalism, therefore, is inescapably touched by disconnection from home and the Cartesian, linear thought pattern. We have seen these two factors clearly demonstrated in previous discussion of the critical issues around the role of science in botanical medicine and holistic herbalism and the plight of endangered medicinals. Unavoidably, American herbalism is a product of American culture. David Milbradt jokes about this.

> Now, herbalism in America is as deep as it is wide. Well, actually it is wider than it is deep [laughs]. We could use some depth of tradition. It is not our fault that our roots have been scorched and tortured and cut off and transplanted and retransplanted again. But there it is. We are trying to get our roots back in the ground again and see where we are. (interview, 1999)

The process of setting down roots is a slow one, but it is essential that the American herbal community be active rather than passive in this process. It is our responsibility to the wanting American public to demonstrate this sustainable connection. As an herbalist once told me, if you know even a little bit more than everyone else, it is your obligation to share that information (Gladstar, interview, 1999).

Some may wonder how I came to the conclusion that herbalism holds only a seed of an emerging spirituality. From my experience in herbalism and my observations of the community in action at herb classes and at conferences as well as during my interviews, I have seen a sense or desire for spiritual connection, but I seldom find a clearly articulated, deeply rooted idea of what that means in practical terms. It is not that I feel there is no spirituality, but rather a rootless spirituality. If the herbal community wants to be relevant to today's society and address its needs, we must enrich our soil, so to speak, and ready it to receive our precious seed of spirituality. Then the sun and rain of the younger herbalists, those new to the field bursting with ideas, will encourage that seed to grow into a tall, sturdy tree for all to climb on or sit under. Finally, the seed of that tree may be carried forth on the air, spreading the wisdom, knowledge, and connectedness of herbalism far and wide.

THE "BORROWING" OF AMERICAN HERBALISM

One aspect of spirituality in American herbalism that needs to be considered when deepening the practice is its essentially borrowed nature. As mentioned, American culture has been and is continuing to be formed by myriad cultural groups; American culture is constantly building and rebuilding its identity; it is the "melting pot," a combination mishmash. American herbalism in many ways mirrors this practice of ongoing restructuring. Songs and ceremony are borrowed from Native American traditions, Meso-American shamanic ways, East Asian Buddhist religion, and Celtic and Druidic lore, to name only a few. This is by necessity. In the beginning, there was nothing. Everything had to be borrowed or created from scratch. As David Hoffmann exclaimed, "We made it all up" (interview, 1998). Although each borrowed ceremony or way of being gives a form or an idea of what to do or how to do it, its meaning does not reach as deep as is meant. The meaning of these ceremonies and prayers and songs, the heart or essence, the power, is held within the cultural context from which they have been extracted. In a 2001 interview, Martin Prechtel, author of an autobiographical series on his experiences living with a Guatemalan village, touches on this point. He says that in writing about the Mayan ceremonies in which he was involved, he often left out details, to the perturbation of his American audience in order to protect the Mayan culture from being co-opted. The songs and prayers are not ingredients in a recipe; there is something beyond the mechanistic application of the ritual. The essential piece, which is beyond the "cookbook" aspect, is embedded within the culture.

Let us examine the analogy of the isolated constituent versus the whole plant to further illustrate the possible pitfalls here. In the same way that holistic herbalists would say a plant is not simply a "carrier" of a certain constituent, a song is not the "carrier" of the ritual power. The constituent is a part of the whole; the song an expression of a culture. Taken separately, the isolated constituent becomes much more akin to a synthesized drug than an herb. Taken out of cultural context, the song is merely words and a tune. The stories that explain it, that make it relevant, the cultural implicit understanding that frames it, have been removed. The constituent or the song exists as part of a whole, dynamic system that holds the healing or the spirituality. It is inaccurate to observe the active constituent of a plant and say it does the healing exclusively. In fact, the whole plant acts on the whole human body and the "active" constituent is merely a part of the action.

In this analogy, it would follow that such a Cartesian, isolationist attitude toward spirituality is also skewed. The song taken out of the original cultural context is just a song, just a part of the whole. Until its stories and myths are restored, it does not have the power of its original intention. Just as the interplay between herb, herbalist, and client is essential to the practice of holistic herbalism, the relationship between song, singer, and culture is key to spiritual practice.

Furthermore, herbalists know that certain plants have constituents whose actions are too strong for the human body to withstand in isolation. However, as part of a whole-plant complex, the toxicity of the one constituent is tempered with that of others. Plants containing cyanogenic glycosides are a good example (Mills, 1991). Botanicals such as wild cherry bark *(Prunus serotina)*, hawthorn *(Crataegus oxycantha)*, and yarrow *(Achillea millefolium)* all possess cyanogenic glycosides, which act on the parasympathetic nervous system and provide antispasmodic and sedative actions. These herbs have been safely used for hundreds of years, yet if the cyanogenic glycosides were extracted and administered, the patient would die from cyanide poisoning. Cyanogenic glycosides break down into hydrogen cyanide, which interferes with the ability of oxygen to bind to hemoglobin, thus starving the cells of oxygen (Cheeke, 1998).

The whole plant, however, confers a useful therapeutic effect. In this way, a potentially lethal constituent can be ingested to great benefit due to the synergy of the whole. I make this point to illustrate the pitfalls involved with taking a song or ritual out of its context. A Native American sweat lodge at an herbal gathering of nonnatives in this view is similar to a standardized extract of a plant—a constituent in isolation. If herbalism seeks true holism, as I believe it does, then I ask that this point be given due consideration and not be reactionarily cast aside by those offended by my criticism of the "borrowing" in American herbalism.

GIVING BACK

Now, borrowing is one thing, but if American herbalism wants to participate in this practice, it must also give back to the cultures from which it borrowed, creating dynamic two-way relationships. In an unpublished 1996 interview with Dr. Tieraona Low Dog, a Lakota herbalist and medical doctor now practicing in New Mexico, she explained the anger she feels when nonnatives "borrow" native traditions but fail to respect the living tradition. Her passion drove home the point that one cannot use a Native American sweat lodge ceremony, for example, as a spiritual practice without participating in a reciprocal relationship with the people who originated the tradition. If one wants to take part in indigenous traditions one must help the living culture to remain vibrant. How can one appreciate the practices of a people and not appreciate the people themselves? This reciprocity could take a number of forms, such as simply donating winter clothes that will be sent to Native Americans living on reservation lands in North Dakota, as Dr. Low Dog suggests. As herbalists we should be extra conscious of how we can develop positive ties to these rich traditions. A simple system of gracious exchange is a beginning. As nonnative Americans, we appreciate native wisdom and custom; they have wisdom, but often they do not have essential material goods. If we have something to offer, it is our obligation to give. If we expect to be the recipients, we must also give in concrete ways.

In Rosita Arvigo's learning about Mayan spirituality and plant medicine described in her book, *Sastun* (1994), she constantly sought to give something tangible back to the living Mayan culture. She was given a gift of knowledge, and she attempted to return the favor with a gift of her own. As an organizer and spokesperson, Rosita has dedicated herself to preserving traditional Mayan healing knowledge. She has helped in the development of the Belize Association of Traditional Healers and regularly accompanies Belizean healers to America where they speak to herbalists about their tradition and work as healers. She connected with Michael Balick and the New York Botanical Gardens to produce educational videos on the Maya and their healing ways. Rosita has also helped to preserve a large tract of prime rain forest in Belize called the Terra Nova Medicinal Plant Reserve. Although her actions are perhaps larger than what the average herbalist would be able to accomplish, they are described as an example of reciprocity. Rosita protects Mayan knowledge, wisdom, and land as if they were her own. In this way the circle of respect is complete.

THE HERBALIST AND THE SPIRIT: UNIFICATION

Rosita Arvigo's story of her apprenticeship with Don Elijio, while quite simple on the surface, is amazingly complex once one delves further into it. Rosita journeys from North America, a relatively spiritually devoid landscape, to Central America, where a deeply rooted Mayan spiritual tradition still exists. She comes from a family of immigrants of Assyrian and Italian background. Although these cultures have strong traditions, most of them were left behind in the mother country; immigrants come to the United States because they are looking for something different. In that sense, Rosita comes from a place of rootlessness.

While in Belize, Rosita works closely with Don Elijio and follows his traditions and way of life. They awake at dawn, harvest plants with prayers, heal those who come in need, pick the domesticated crops, and perform Mayan rituals together. Rosita, step by step, adopts these prayers and traditions as her own; she begins to practice them without the direct guidance of Don Elijio. As she does so, Rosita became more solidly rooted in the rich soil of Maya spirituality.

Part of the development of a rooted practice of healing, and of life, in Rosita's case, involves a recognition of the unity of knowledge and spirituality, of healer and gods, of interior and exterior forces necessary for successful healing. She writes, "In the Maya world only a gossamer veil separates physical from spiritual" (Arvigo, 1994, p. 37). She understands early on in her journey that the two may easily blend one into the other. Rosita notices that Don Elijio is "more than just a man of the plants" (p. 29). He brings together skill at healing with spiritual adeptness. Rosita herself came to Belize with a bank of healing knowledge only to find her healing powers intensified when she also employed the assistance of the plant spirits and Maya gods.

Throughout *Sastun,* Rosita makes references to the unity involved with healing, showing that one must say prayers so that the plant spirits will *help* in the healing. The spirits follow the harvester home and potentiate the healing power of the plant and the curandero. Although the spirits help with the healing, at no point is there disregard for the role of the healer. The healer undeniably has knowledge and skill. Don Elijio says he has it all up here, pointing to his head. He possesses great understanding of the human body and spirit but is greatly assisted by the spiritual forces.

Such sentiment is not isolated in the indigenous world. Curanderos, granny healers, and midwives as well as shamans throughout the world suggest the same phenomenon. In *Messengers of the Gods* (1995), a documentary video concerning traditional healing in Central America, Hortense Robinson, a midwife of Mexican origin, speaks about listening to dreams and the voices of the gods as they come through the dream structure. She says that one will learn more quickly if one can tune in to the messages from the spirit world. She does not say that one cannot learn without listening; it will simply quicken the pace of learning.

Eliot Cowan, practitioner of plant spirit medicine and author of a book by that title, interviewed shamans and spirit healers for his publication. Throughout the transcripts are references that the plants indeed have spirits and those spirits are catalysts or assistants in the healing. In the volume, Enrique Salmon, an ethnobotanist and shaman, says a healer must sing to the plants as they are harvested "to get the medicine out of the plant to help the patient more" (1995, p. 132). The song catalyzes a reaction that allows the spirit and the healer to interact, thus providing stronger medicine to the patient. Again, in another interview, Siri Gian Singh Khalsa speaks of a similar thread. He mentions that each plant must be addressed in a specific way. He says, "Something had to be done over the herbs in order to activate them and to make the herb an ally to the patient" (1995, p. 160). Another healer follows that same vein saying, "I find that the main thing is to get the permission of the spirit that is going to help me to help others" (1995, p. 168).

Such statements begin to show that throughout the indigenous world, healers speak similarly about plant medicine as a collaboration among the healer, the plants, and the spirits. For the American herbalist, as well as the indigenous herbalist, there must be unification between the power of the herbalist and the power of the spirit. To credit either as the sole originator of healing would be inaccurate and incomplete; such a view is distinctly linear and dualistic. A relationship must be considered; a system must be recognized. Within the paradigm of the indigenous healer, the herbalist is the holder of knowledge for other humans. She or he has the understanding and wisdom to employ and enchant the plant spirits and to know which plants might be helpful for which condition.

Throughout this book, systems and relationships are discussed. It is through a systems view of the world, an acknowledgment of the constructive chaos-based feedback loops that amalgamate our reality, that we can begin to understand how to maximize our usefulness on this planet and be fully human. In the concluding chapter of *The Web of Life* (1996), Fritjof Capra notes the following:

> The origin of our dilemma lies in our tendency to create the abstractions of separate objects, including a separate self. . . . To overcome our Cartesian anxiety, we need to think systemically, shifting our conceptual focus from objects to relationships. . . . The belief that all these fragments—in ourselves, in our environment, and in our society—are really separate has alienated us from nature and from our fellow human beings and thus has diminished us. To regain our full humanity, we have to regain our experience of connectedness with the entire web of life. (pp. 295-296)

It is through recognition of this interconnectedness that we as herbalists can reach the healing potential for which we are meant. Annie McCleary spoke eloquently to this point in our discussion.

> When I work with plants, I work with the vital energy that is there. I speak to the plants and give them my energy; they speak back to me and give me their energy. We're in communion; we're working together. They know that we replant. They know that the ecosystem is not being damaged by pesticides or herbicides. We grow our herbs organically. We're simply going with the harmony of the universe. If you want to restore harmony to the human body, you need to restore it with something that is itself in harmony. You can't restore harmony with something that's out of harmony. (interview, 1999)

Annie keyed into the reciprocity, the relationship, in a meaningful way. So often we see only one source of wisdom. Biomedical science and much of botanical medicine sees doctors or the medicine person as the healer with the "power" to heal; some brands of spirit medicine see spirit forces as the sole doers of the healing; herbalism sometimes sees the body of the individual as the only active participant in healing. As visioning herbalists, we must strive to combine these views into a more comprehensive and relevant system. We must bridge the diverse healing paradigms, uniting all of these powers and catalyzing a synergistic reaction between them. It takes each part of the system to liberate a fully realized potential. If herbalists are to bring healing to the self, the community and the environment, if they are to fully engage in the herb-herbalist-client relationship, they must bring the unification of the healer and the spirit to the forefront in the field and understand the necessity of each supportive element in the healing process.

As mentioned, Martin Prechtel writes in his memoirs about the "indigenous" soul in each of us. Poetically, he describes his own story of remembering and peeling back the layers to expose this facet of his humanness. He succeeds, as does Rosita Arvigo, in attaining a deeply relevant spirituality within life and healing. Both clearly demonstrate the ability of any person, no matter what his or her background or course of study, to reach this point of interconnectedness. If American herbalists heed their advice and example, as well as those offered by participants in this study, and clarify the process and spiritual ecology of the field, the potential for American herbalism as a catalyst for dramatic mainstream paradigm shifting will be immeasurable.

Over time, a solid identity for American herbalism will blossom. This will be formed partly by people who immerse themselves in traditional cultures, and return to transmit that complete cultural knowledge to other American herbalists. The strongest part of the American herbalist identity, however, will come from those working on the grassroots and community level attempting to develop ever-stronger bonds between themselves and the plants and then communicate that outward. Kate Gilday spoke of these beginnings.

> We are growing into our birthright in understanding the connection between ourselves and the plant; it is incredibly sophisticated. It is incredibly deep and detailed. It is like a web. Everything touches everything else. Those of us who work with the wild plants in any way, whether we are growing them or harvesting them or using them in our practice, we are getting to know that complexity. It is the beauty of how the herbs work in the body. (interview, 1999)

To accomplish these goals, it will take more and more herbalists finding grounded ways in which to talk about plant allies and the possibility of communicating with plants and having a reciprocal relationship with plants. Leslie Gardner introduces this idea to students as they come through the California School of Herbal Studies.

> One of the things that I often say is that you can begin to trust your intuition. And begin to recognize that it might not always be coming from you. The things that you think are your ideas—oftentimes it is just because we are trained to think that way. In our Judeo-Christian framework, we have been trained to think in a dominator model with a patriarchal system over it. We don't think of ourselves so much in concert with the plants anymore, as I think we did back in the old, superstitious days of the Middle Ages. There was a lot of superstition, not science. There wasn't anything we could pin it on. It was all very mystical. Those same messages still come through the plants. It is just that we don't interpret them in the same way. One of the things that I encourage people to do is to recognize or to open themselves up to that communication. To me, it is pretty essential. . . . That sense of opening up to something greater than what you are. It doesn't matter what you call it. Words get in our way. But spirit is something that a lot of people will accept. . . . It is a matter of allowing people to recognize that they can flow into something that is out there. That they don't have to dominate the plants. They can be in harmony with the plants. The plants are our allies. There is a partnership going on. (interview, 1999)

With clarity and vision, grounded unification of the healer with spirit, deep understanding of herb-herbalist-client dynamics, and community networking (herbalists talking and working together on common goals), the potential for American herbalism expands exponentially. Now is the time for American herbalists to sink their roots deep into the soil of the American culture, deep enough that the drought and harshness of an Earth- and spirit-alienated culture does not bake or wither them. With a solid rootstock established, we can begin reaching tall branches to the sky, providing shade for those who need it and a stepladder for those who want to reach higher.

PART II:
THE THREADS—INTERVIEWS
WITH TWENTY AMERICAN
HERBALISTS

Chapter 7

Laura Batcha

Laura Batcha is a folkloric herbalist and medicine maker. In the early 1990s, she began a farmer's market business centered around bioregional herbal extracts that she turned into a national supplier of high-quality herbal tinctures and innovative formulas. At Green Mountain Herbs, Laura supervised production and formulated new products. She also coordinated with Green Mountain's organic farm and other local producers to obtain the freshest, most potent raw materials for her extracts. As Laura mentions in the interview, Green Mountain Herbs was bought by Tom's of Maine. Laura is currently the product development/education team leader at Tom's of Maine. She continues to formulate, garden, and occasionally teach.

I spoke with Laura at the Green Mountain Herbs headquarters in Saxons River, Vermont, on July 26, 1999. I asked her to share her experiences of steering a rapidly growing herbal business through the herbal renaissance.

GREEN MOUNTAIN HERBS (GMH)

I started in the business not with the intention of being in business. That sort of happened on its own. I got started with Green Mountain as a business by just loving to make extracts, working a lot with plants and studying plants. The thing that I really connected with first off was making medicines from what I was gathering and growing in the garden. I started making different extracts and syrups. Every time I would learn about a new plant I would want to extract it. My house quickly filled up with tinctures. Neighbors would stop by to get herbs from me. I would write handwritten labels, and I would tape them on the jars and send them out with that. Then neighbors and people in the community would come by at all different times of day and night. It was always catching me off guard because I didn't actually have products. To create a little order in my life, I decided to go to the farmer's market on Saturdays. My kids and I would pack up a little case. We bottled the extracts and we mimeographed labels. We would bring them down to farmer's market on Saturday; that way people could come and meet me there rather than coming to the house. That is really how it got started.

Then local stores approached me at the farmer's market, wanting to carry the herbs because people liked using those herbs rather than what they were [selling] in the stores. At that point, I created nicer labels and went from there.

The bulk extracts that we sell came from the farmer's market as well. I would encourage people to bring the bottle back in a couple of weeks when it was empty. I would bring stock jars and refill for them at the market with a funnel. As I stopped going to the farmer's market and was really just selling to retailers, [the bulk extract displays] stood in place of me being at the farmer's market and refilling the bottles myself. People could bring their bottles back and refill them at the store.

Since then we started selling to stores in Vermont and then New England. Within a couple of years we went to one of the national trade shows on the West Coast. We started selling across the country.

Just this last spring, I sold the company—or merged the company with a larger business. Now I just do herb stuff again and don't run the business. I have sort of come around the horn with it. So, right now, Green Mountain is officially owned by Tom's of Maine. That is new this spring. Now my work is primarily back again to developing products and formulating and educating around herbs and quality control over the extraction. I am back doing the same things I was doing when I got started. For me the interim period was learning about commerce. Four years of commerce intensity, and now I am back to the herbs again.

RUNNING A HEART-CENTERED BUSINESS ON THE NATIONAL LEVEL

I never felt that the heart was slipping away [from my herbal practice] and that national commerce was taking over. I did feel like I did not want to learn what it would take to compete on the commerce level. I don't think I ever felt that the heart was slipping away because we were never under the pressure of having venture capitalists or any investors in the company. If we didn't achieve our goals [we didn't have to answer to anyone but ourselves]. From that perspective, we were able to maintain our center. Through the time that the business grew, the thing that we gave on was we fell short of being able to function in the marketplace. For example, our first year in distribution we probably "out-of-stocked" our distributors twenty percent of the time. It really got the distributors angry, but rather than changing the way we did things and giving on the approach we were taking, the commerce end had to give. That was the way we drew the line as we grew. The marketplace is increasingly competitive. They will only take that from you for so long.

The reason that we partnered with Tom's of Maine was because I felt that it was an opportunity to see the company continue to grow and be healthy and maintain its value center. Tom's is so well known for being a value-centered company. The two companies had a lot in common in that way. When the opportunity presented itself, I felt that I could draw on their experience of the last twenty years. They have experimented with how to maintain values in business. They are the best, in my opinion, in the country at doing that. They are very successful and very profitable, but they have core values that they don't give on in terms of how they do business. For me, it was an opportunity to bring that structure into our company. It is really difficult to do. Tom's started doing business at a time when the industry wasn't what it is now. It is more difficult to do it now than when they were getting established. One of the main reasons why I made the transition was so that I could focus on the core values of the company rather than on the details of commerce. I am more suited to be the caretaker of that than I am to maintain bottom lines.

The problem for us at Green Mountain was that we were the opposite of most companies in that I have no business training. I have no economic background. I have never taken a business course. I have no background in that at all. So, it is amazing we actually made it as far as we did just by winging it. We had a good product and a good idea. People were drawn to the energy of the company. We got lucky in terms of that, but you have to look at gross margins and those types of things to maintain the health of a system, you can't ignore those. I not only didn't have the skills to do that, I don't really have the desire to do that either. That was something that didn't even come into play at Green Mountain. It is just amazing that we made it as far as we did without paying serious attention [to the business aspect]. I don't think we ever really came up against losing the heart as much as we were aware that we weren't operating the way everybody else was, and it was starting to get difficult.

THE BOOM IN NATURAL PRODUCTS

Natural products are more mainstream now than they were fifteen years ago, obviously [Eisenberg et al., 1998]. That means that there are different consumer requirements. It used to be that you could have a good product and be good folks and be funky and get out there and just on the value of what you were doing, you could succeed. I don't see that as really being the case anymore. You have a to be a savvy marketer. In terms of your branding, you have to be visually very contemporary. You have to compete against really large companies that are putting a lot of money into marketing. Green Mountain is a very small company. Two years ago we had half a million dollars in sales. Our competitors were twenty million up to one hundred million-dollar-a-year companies. You have to compete. When they walk up to a shelf, the consumer can't tell the difference between the two companies. When you are at a trade show, your buyer can't tell the difference other than the size of your booth. You have to function at that level. The visual standard for products has also really escalated. No "funky" anymore; it just doesn't fly. So, that is one thing, you have to compete with bigger companies and you cannot appear any different. It just doesn't work.

The other thing is the distribution consolidation that has happened in our industry. There are really two primary natural product distributors that have gone national and bought up other ones.[1] To go direct, you have to have a strong infrastructure and be a large-enough company to sustain that, to direct sell to stores. Otherwise, you have to play with those big distributors. You pay the same for advertising as the hundred-million-dollar-a-year company. The distributors want the same ad contract. It is more expensive for a small business in terms of that.

Some of the other things that come into play are new regulations, particularly in terms of the herb industry. I am finding that there is a gray zone for our good manufacturing practices (GMP) [FDA, 2002] where you are exempt. If you have less than one hundred employees and you sell less than ten thousand units of any given item, you don't have to comply with the DSHEA [Dietary Supplement Health and Education Act—as outlined by the FDA] laws for that particular product that you sell [CFSAN, 1999]. But because buyers and store retailers and consumers are aware of the new GMPs—there is a lot of media coverage out there—they expect you to be compliant, even if you have an exemption. It is incredibly costly for small companies to be compliant with the way the industry is going.

CURRENT FEDERAL REGULATIONS

I am in support of the new FDA regulations. They make sense for large companies. They are already putting crap in the marketplace. Hopefully, this will ensure that they have to say that it is crap [laughs]. They have to disclose it on the label at the very least. I am in support of it because it is the only way that really big companies can be held accountable. Often times with these big companies, there isn't even someone in the building who knows anything about plants. Without having those regulations, it would be a free-for-all for consumers, and they would be taking products that are adulterated, and nobody would ever know.

For small, hands-on companies, it is not as important because they tend to be run by skilled people. In terms of consumer protection, it isn't as important, but somehow that hasn't translated to the consumer. It crossed somewhere in there. Consumers have more confidence in a large company than they do in a small company because the labeling appears to be in compliance. In my opinion, those are riskier products for people to be taking than from really small manufacturers simply because of the skill level that is involved in production.

I think that the new regs are one thing, and the other thing is the trend toward the popularization of herbs. This has brought them out of the realm of traditional products and more in line with pharmacognosy and the top fifteen herbs. The new literature that is out on plants, on one hand, is good because we do live in a scientific age, but it creates a burden for small manufacturers to keep up with the testing involved. Green Mountain Herbs, as an independent company, could never have afforded lot-to-lot microbial testing, and lot-to-lot HPLCs [high-performance liquid chromtography, an analytical method for identifying the presence of certain plant constituents] on our products whereas a larger company like Tom's of Maine [has] an in-house R&D [research and development] department. They routinely do all those tests.

Those tests are not required by anybody in terms of the GMPs except that that information is out there about plants, and that is what buyers want; that is what consumers want. We are always moving toward the future, whether we like it or not. [Scientific assays are] the future of how we are going to be interpreting plants and their effectiveness. Hopefully that will not be the only way we do it. I think that the trend is much more balanced than it looked like it would be even two years ago. I am pleased about that. I think we are not going to go off the standardized deep end in this country. I think we have pulled back from the cliff on that. It is much more reasonable.

SMALL COMPANIES AND CONSUMER EXPECTATIONS

We are working really hard to get the message out: that small, traditionally based companies produce good, effective products. Also [we are] learning from Europe and the approaches there. That has brought in a different, more scientific perspective. I think we are going to end up somewhere in the middle. I don't think we are in real danger of completely losing the tradition, but that is the way new information will come out about plants. Smaller companies do not have the resources to justify and rationalize their products by [scientific assays]. It is like most things in America now. The gap between the small and the big is really polarizing. You don't have to be concerned with those things if you are a small, regional-based company or if you sell direct to consumers, retail mail order, that kind of thing. As long as you stay on the periphery of the industry, it isn't really that much of an issue. There is still a huge opportunity for real small [companies], but when you get in that in-between size, that is where the pressure really exists.

Small companies just starting up will have to be more savvy than you had to be ten years ago. You have to get a niche and reach an audience. That is the way that it is. I think that there is more room on the small, regionally based level than for an in-between size. A small company trying to compete in the industry is at much more risk than a small company that is choosing to stay on the periphery.

THE MISSION OF GMH

It is our mission to support traditional herbalism using the variety of the European and North American pharmacopoeia. We have to adhere to that. The company is built on the foundation that we won't ever reduce our product line to only the top fifteen clinically sold herbs. My training is as a traditional herbalist. I learned through apprenticeship and through experimentation even before there were all of these new herb schools that are out there now. The formulations come from my experience with using the plants.

I formulate based on a couple of things. The first is herbal actions, which is the lost art of herbalism as we go into pharmacognosy. It is very different from the chemical constituent approach. That is a real pillar of traditional herbalism that has to stay active. You can't always rationalize it in terms of pharmacognosy. We have words like soporific, something that makes you sleepy. From an herbal action perspective, if the herb makes you sleepy, it makes you sleepy. It doesn't matter what the mechanism is that is doing that. You are formulating based on the fact that it is a soporific. You are not formulating based on a mechanism. That allows you to draw in a broad range of plants when you are approaching something. I am not sure if that is the best example I could come up with, but there is a difference between looking at the biological activity [or mechanism] of a component of a plant and the herbal action. The herbal actions allow you to really narrow in on the unique personality of the plant. If you approach a plant, say chamomile, and you list the herbal actions—it is an anti-inflammatory, an antispasmodic with an affinity for the digestive system with secondary bitter qualities—that tells you much more about the plant in terms of how to actually use it in a formula than if you just looked at the components.

I also use the energetics of the herb. I would liken that to cooking and tasting. I formulate a lot by tasting. You can tell you are on to something when they really work together. I really like the quirkier plants so I tend to not be afraid to use them. You stick with a few solid ones that a consumer will recognize, but I always then bring in what I call the unsung heroes of herbalism: balmony, nasturtium seeds. A lot of different plants. I really like to bring those in because I don't want to see them disappear from herbalism.

The herbs that we tend to use are fairly common and easy to grow just because being a vertically oriented company—we grow the plant, make the extract, sell it, market it—so the easier it is to produce, the more likely it is we will succeed. I tend to use things that are not that hard to grow [laughs]. If it is really hard to get ahold of, I will shy away from it. It is just not that easy to complete your mission of being your own supplier of the raw materials. If I have a choice between two herbs of comparable action to put into a formula, and one is easy to grow and the other you can't really cultivate and there is not that much of it, we always go with the easy to cultivate.

ENDANGERED PLANTS

When I started studying plants, my approach was that if it were a marginalized plant, I would simply not learn about it in terms of its therapeutic benefit. That way you don't end up in a position of feeling like there is no substitute. In our line, we didn't even start manufacturing a goldenseal extract until two years ago when we got a cultivated supply. I have been working with plants for fifteen years, and I had never used goldenseal before. It had always been endangered in Vermont and nobody was growing it back then. I simply didn't learn about it. I didn't have any notes on it. I didn't have a file on it. That way, I was never tempted to use it because I was never in a position to know that much about it. That is my own approach.

CERTIFYING EFFICACY

There are a couple of ways to certify efficacy and it is actually a question we are dealing with right now. Now that we are part of Tom's of Maine, I am doing some reformulation on the products for a relaunch after the acquisition at the end of this year. We are positioning Green Mountain as a complement to some new products that Tom's of Maine is coming out with which are going to be more of the pop herbs, mass-market type products. Then, Green Mountain will be a complementary line of traditional herb products. Those will be set in natural food stores and not in mass-market stores. They won't be in the corner drugstore because that is not what people go to the corner drugstore to look for.

We have been dealing with the FDA and claims and quirky herbs. You can't make a substantiation file on some of the plants that we use. The DSHEA [CFSAN, 1995] doesn't require clinical trials to make a claim. In the language of the law itself, you can make a traditional use claim. The FTC [Federal Trade Commission], however, does require clinical trials to substantiate advertising claims.[2] That is where the push has come from in the industry, that you have to have a clinical [trial] in order to make a claim. The FTC is much more proactive in enforcement than the FDA. The FTC won't allow a claim in advertising or in product literature in catalogs that do not have a clinical trail associated with it. It is the "truth in advertising" clause. That is going to be the police arm even though it is not part of the FDA at all.

What we are doing is sticking with the broad pharmacopoeia in formulations . . . like the echinacea ally has echinacea, nasturtium seeds, and cleavers in it. That formula is fifty percent echinacea, so according to *German Commission E Monographs,* I hit therapeutic dosage of echinacea. I can then make an echinacea claim on a product like that. I don't have to make a claim on the other ingredients. You can legally make a claim on a product if you have an ingredient that is backed with clinical trials, and you hit a certain dosage level in the formula. If the formula was five percent echinacea, I couldn't make an echinacea claim. A lot of the formulas can be rationalized for claims by having lead ingredients that carry clinicals as long as the other herbs you have in there don't have negative safety reports associated with them. I couldn't put comfrey in there and make an echinacea claim because comfrey has a negative safety review. Even if that claim isn't on that ingredient, you cannot have ingredients that have negative safety reviews if you have a claim on something else in the formula.

The other way is to not make claims on a product at all. We do that with a lot of products. You just rely on the fact that there are a lot of good folks out there who are learning about plants and they know what they want. They are going to look for it. At that point, you rely on third-party literature.

We are going to try and see if we can't substantiate some files on some lesser-known plants for claims. Right now, there are fifteen plants that are typical claims plants. Hawthorn, St. John's wort, those are the top ones. There are, if you go through the literature, there are a lot of clinical trials on lesser-known plants. One of our formulas is goldenrod ally for allergies. I was just reviewing the literature on goldenrod and there are three or four clinical trials on goldenrod out of France. That really surprised me. It is just that they are not that popular so people aren't even building files on them. So, there is opportunity to rationalize those traditional ingredients, but you have to dig for the information.

NATURE OF CLINICAL TRIALS

There aren't actually requirements for trials at all. If you look at them, some of the clinicals have as few as seventy-two people in them, which is absurd. We are building this entire fortress around a clinical trial with seventy-two people. It is almost shocking. If you go back and look at the clinical trials that most of these claims are based on, they are not great trials. They are not extensive. In my opinion, it is not really worth hanging your hat on it if it doesn't corroborate traditional use of the herb. For St. John's wort, there are probably three thousand individuals who have participated in clinical trials, but there are some herbs that routinely carry claims in the marketplace based on clinical trials that have as few as seventy-two people participating. You or I could design a study with seventy-two people. The only real requirement is that the study has to be published. You have to have somebody with credentials to publish a study on board. You have to have a PhD signing off on the study with the connections to get it into a journal. It is a university-based network. Even if it is done by private industry, there has to be somebody there that can get the study published.

A lot of the companies are shying away from doing their own clinical trials because you are risking having the trial come back negative. At that point, you are set up and going down the publishing track and cannot afford to publish a negative clinical trial. That is why the same products in the same formula to the same specifications get repeatedly tried. Companies make the determination that the risk involved is greatly diminished because you are using exactly the same thing as the person did before. That is the driving force behind that. You can find seventy-two people to take anything. That is not that hard to do, but you can't risk having it come back negative. There have been a lot of clinicals coming back negative on echinacea recently. It is hard to clinically try echinacea. The research is qualitative to a degree. Did you get worse or did you get better? How sick did you actually get? It is not straightforward. I think the reason you don't see more of that is the fear of the clinical trial coming back negative. You could make a claim on somebody else's trial, but if you have a negative trial associated with your product, you don't look very good.

Then you have to be hooked into some publishing loop in order to get a trial out there. You have to be working with some kind of established organization that will get a trial published. In order to do that, they sign on in advance. They publish the trial regardless of the outcome. You can't do a trial and agree to publish it only if it comes back positive. Say we hooked up with the University of Vermont to do a study of whatever. University of Vermont is going to publish the results of the trial regardless. You can't say, can we hook up with you and only publish if it works.

ADVICE FOR PEOPLE WANTING TO GET INTO THE HERB BUSINESS

People say, "I want to do this. What do you think?" In any entrepreneurial venture you have to have a distinctive idea that you want to put out in the marketplace. That is limiting because the marketplace is saturated. You have to be innovative. That is one thing.

You really need to have a clear vision of what you want to do. Don't be afraid to bring skilled people on board early who have expertise that you don't have, like somebody who was trained in marketing, somebody who knows how to evaluate the economic feasibility of what you are trying to do, somebody who understands the science of herbal medicine. Not that those things have to drive decision making; we don't have to buy into reductionist scientific evaluation of herbs, but we can't be completely ignorant of what the scientific community is doing. You have to be as up to speed as they are but have a different paradigm that you are interpreting the information from. We can no longer say, "I don't really care for that" and not pay attention at all. It doesn't mean it has to drive your decision making, but you can no longer afford to just not deal with that. If you don't want to do it, then you have to have somebody there that can do it, who can provide you with the information that you need.

If you want to truly be a small, community-based company, and you don't want to get into industry itself, I think that there is plenty of opportunity, and there always will be opportunity for people to do that. Absolutely. There are a million variations, and there are a lot of people out there spending money on herb products. In terms of getting into the actual industry of herbs in America, I think it is going to be tough for people. The prevailing notion out there in the industry when you go to call on a store is: I don't need another herb line. Please just don't even come through my door. There is such a proliferation. There are newcomers into it also, like Tom's of Maine. Everybody is selling herb products now. It is the segment in natural products that has got the highest growth rate. We are a growth-driven economy. Everybody wants to participate in that. Then, there are pharmaceutical companies coming in as well to get on board for the mass-market growth.

In terms of the industry itself, I would only be positive with somebody [starting a business] who I thought was so completely consumed by his or her vision of something that had to be created. I think that is really powerful, and you could do anything if it is really your driving purpose. If it were a rational, intellectual decision to do it, I would discourage somebody from doing it because it is really tough. It has even changed in the short time that I have been in the industry. I would not want to try to do what I did even five years ago right now. I would probably choose a different route to take. Unless you had money. You would have to secure serious financing from somewhere to do it. You would still have to be innovative, and you would still have to be driven. You can't get by anymore without all of those things in line. We may be the last of one of the few herb companies that can hobble into the marketplace without any significant funding and somehow hack out a little area for ourselves. It would be very difficult to do now.

EDUCATING CONSUMERS ABOUT HEART-CENTERED HERBAL PRODUCTS

Sometimes I think we don't educate very well [laughs]. It is hard because there are real limits. There are legal limits on education right now with herbs. You really have to educate around quality and safety more than you do around use. Everybody is educating around the quality of the products because you don't have legal restrictions on that. You are talking about using high-quality raw material and organics and techniques. Most of the education is geared toward that. We still do a third-party pamphlet that has my name on it that explains how to use all of the formulas. That is on the edge, legally.

FTC regulations discuss how closely you can display that material to your products, and assumptive association according to the FTC is enough. The FTC has gotten much more involved in the last year than they did before.

There are a lot of people that know about plants these days. There are a lot of educated consumers. The thing we rely on most is that there is a movement in this country of people learning about plants. They are reading, they are studying, and they are taking classes. For a company like Green Mountain, that is the core people. Our products don't really appeal to the newest wave of entrants into the herb market. If you were to buy gingko or echinacea for the first time, you wouldn't buy from us probably. We wouldn't be pitching you. You are not our audience. Our audience is a much more self-educated consumer. Those are the people who like our products.

Chapter 8

Richo Cech

Richo Cech is fascinated by the challenge of unlocking nature's secrets of growing wild medicinal plants and has written many publications on medicinal herb cultivation, including his newest books, *Making Plant Medicine* (2000) and *Growing At-Risk Medicinal Herbs* (2002). For many years, he worked as herbalist and research and development person for Herb Pharm, an Oregon-based manufacturer of liquid herbal extracts. Currently, he runs a medicinal herb seed company called Horizon Herbs. He lives with his family on a small farm at the base of the Siskiyou Mountains in southern Oregon, where they grow over 300 varieties of medicinal plants for seed production. Richo serves on the executive board of United Plant Savers. He believes that the organic cultivation of medicinal plants provides a necessary alternative to the harvest of precious plant resources from the wilds.

Richo and I spoke at the United Plant Savers Member Conference, June 27, 1999, held at their first botanical sanctuary in southeastern Ohio.

FACTORS CONTRIBUTING TO AT-RISK PLANTS

I think the current situation with at-risk plants comes from a lack of sufficient habitat, basically. I think the main reason that there is less and less plant life available is because there is more and more cement in the world [laughs]. I recently came from Kentucky where I have interest in a farm, and we have a large area of wildlands surrounding this certified organic farm. It is over a hundred acres and contains twenty acres of prime habitat, that I have roamed for the last seven years, that has literally hundreds of different plant species of interest to herbalists. When I came back this year, it was bulldozed without me even knowing it. Twenty acres of goldenseal and ginseng and several different varieties of wild yam and of course all the supportive, overstory trees were simply gone. That is about development. The land is being purchased and broken up into lots. People are buying lots there because everybody wants to get away from the cities, and it is close enough to the major urban areas that people can commute. I think that is really the biggest problem.

I don't really think that the numbers of at-risk plants come from a lack of consciousness on the part of wildcrafters. Many wildcrafters, especially bioregional wildcrafters, have a consciousness about what medicine really means. They understand that when you pick the plants with a good feeling and with respect for the plants and respect for the Earth, then it produces the desired effects for whoever takes it. You know, you have to instill that kind of consciousness and care in the medicine from the very beginning otherwise I feel that it doesn't really work. That ties in not only on an ethereal level, but also on a practical level. Preparing high-quality medicine has as much to do with how careful you are with the drying and the preparation of the plant material, as well as the kind of plant parts that you choose and the health of the plants that you choose to pick. So, it has not only an ethereal side, but also a very practical side as to why that functions in that way.

There is, however, another faction of wildcrafters who are working for big companies. We were just getting some information yesterday from a plant conservation researcher who said that she had watched the Pepsi trucks unloading *Chimaphila umbellata* [pipsissewa] by the bale.[1] We all know that pipsissewa is something which, like most wild plants, is locally common or locally available, but when you go to the fringes of its habitat, it is much more susceptible to any picking. *Chimaphila* is susceptible to picking even if you are in the midst of a big stand. Two whorls have been picked off the top of a plant, which usually has three or four whorls, and we have seen that after four years there is only half an inch of new growth. It is extremely slow growing, and even if someone is out there respectfully taking a quarter of the patch, it is a scar that is going to be visible for a decade or more. When a big company goes in to pick pipsissewa by the truckload, you know they can't be giving it that kind of care. They are going to be pulling the plants up by the roots, swishing them off in a tub that is next to a truck somewhere, laying them out to dry, and then baling them up and taking them out for extraction for flavoring purposes.

I don't think that it is the responsibility of the wildcrafter, or it is the karma of the wildcrafter population, that we are running out of wild plants. I think that there is a certain dent that collection for medicine puts in wild populations, and that this can be done sustainably on a bioregional level with people making medicine for their family and friends in need. I think there is a bigger dent put in it by certain larger companies and corporations coming in and harvesting on a large scale.

The loss of habitat on a large scale is the main problem that UpS sees, and that is why we feel like one of the real good solutions, besides the obvious one of cultivating plants, is to make nature preserves anywhere we can [especially in those places] where we see the centers of plant diversity and what are known as the gene centers for these various species. For instance, right now, we are sitting in the Ohio River Valley. This is the gene center for goldenseal in the United States. This is where the major populations have been. If the plant has grown out from somewhere and populated other areas in, for instance, the Midwest—the Missouri Ozarks—or further north into Canada, it was basically moving from here out to there. These are the areas that are most significant in terms of preservation so that we can be sure that we have wild germplasm to develop into our cultivation scenarios. Another example might be preserving prime osha habitat in Colorado. There are a lot of new homes that are being built up in the mountains around Boulder, and the places that they choose to put the houses are almost always prime osha habitat. Osha gets dug out and driveways get put in. Obviously, there are places in the Rocky Mountains in the front range where osha is very common, but there also areas where it has been dug out completely, or where it has disappeared because people have changed the habitat sufficiently so that it cannot grow there. If you view the plants as part of a whole environment, a whole wild environment, then you realize that all it takes is for the environment to be sufficiently changed and the plants will disappear.

CRITERIA FOR LISTING ON THE UpS AT-RISK LIST

The criterion that we used was first and foremost that the plant is economically significant. We realized that, for instance, some of the gentians that grow in the mountains are extremely rare. Some of them are on endangered species lists in certain states, but there are very few herbalists that are going to go out and pick wild gentian. The amount harvested is not really significant, and you don't see products on the market with wild American gentian root as one of the ingredients. So, gentian didn't go on the list. The False unicorn, *Chamaelirium luteum,* is something that is used in medicine, however. It is used here; it is used in Europe. It is a slow-growing plant that is difficult to propagate and it is economically significant. For those reasons, that one goes on the list.

The other criteria we used was basically that it is getting harder and harder to find these plants in all the different niches of their habitat, and all the different areas where their habitat occurs. *Echinacea angustifolia* is going to be on the list because there are plenty of places in Kansas where it used to be part of the wild prairie, and you can't find it anymore. There are a few little ones sprouting from pieces of the root that were left in the ground when people dug them out because you can't get the whole root, but that is nothing compared to the beautiful resource that was there. It is every American's birthright to be able to go into the wild prairies and see medicine plants growing, and if they are being taken out at an untoward rate which isn't sustainable, it makes sense to let people know that this is happening and try to develop positive programs that give the herbs a rest.

ROLE OF MANUFACTURING IN THE ENDANGERED-PLANT CRISIS

All manufacturers need to bear responsibility for the sustainability of the raw materials that they use. There is a rising awareness of the need to cultivate plants. This is coming in part from the populace. They are becoming more and more aware that it is good to examine the label and see if an at-risk plant is being taken from the wild or if it is being cultivated. That has a strong effect on manufacturers because they are looking very carefully at their audience and using that to clue them as to where they need to go in order to increase sales.

Increased awareness is also coming from the manufacturers themselves. As these plants are picked out at a greater and greater rate, and the habitat is disappearing at the rate of twenty-four hundred acres per day in the United States [Liebmann, 1997], there is simply less wild-plant material out there. The laws of supply and demand kick in, and the herbs become more expensive to get. Also, the available roots become smaller because it is harder to find the big roots, which are going to weigh up heavy. It is a good indication to manufacturers when they start looking at an incoming shipment, and they see roots that are smaller than usual. They are realizing that the population is being diminished because there is no reason why a harvester would be going after the little stuff at four dollars an hour when they could be getting the big stuff at twelve dollars an hour, except for the fact that the big stuff isn't there anymore.

The government is becoming increasingly aware of these problems on all kinds of different levels. Fish and wildlife is the main government entity that is responsible for putting the teeth in any environmental laws that affect the harvesting of plants out of federal grounds. Then there are environmental organizations that come into play here. TRAFFIC International and United Plant Savers are working in cooperation with other conservation organizations to see what we can do to raise public awareness. It is through public awareness and the willingness to read labels and determine what the true source of the plants is that the main revolution is going to occur. It has to come from a grassroots level.

RATIONALE AND EFFECT OF LISTING GOLDENSEAL ON CITES—*APPENDIX II*

CITES (Convention on International Trade in Endangered Species) encourages the sustainable use of endangered species through the regulation of exportation. Goldenseal was added to the CITES II list in 1997, sparking industry-wide discussion about the use of the plant in commercial preparations. Further discussion about the pros and cons of this listing according to United Plant Savers is included in Anaheim (1998).

In my opinion, it was a good move to add goldenseal to the CITES list because it brought medicinal plant conservation to the forefront. It could have been any of a number of plants that were used. Goldenseal didn't have to be the one that we utilized, but it is a good one because it is a strictly native American medicinal plant. It doesn't occur on any other continent in the world, and the harvest levels are substantial. In 1998, two hundred sixty thousand pounds of dry goldenseal was harvested out of America and only sixty-five hundred pounds of it came from cultivation. The predominance is in wild harvesting. All you have to do is talk to wildcrafters to know that there are a lot of goldenseal places that have been picked out already. It is clearly a problem. Goldenseal is poached from federal lands and in places where it is not legal to pick. It is poached from private land also at a disturbing rate. Landowners around here used to try to keep people out who were going after ginseng. Then, there came a period where people weren't walking on their lands anymore because the ginseng was all picked out. Now they are seeing people coming in again because the price of goldenseal is so high that they are starting to get the goldenseal out.

Listing goldenseal on CITES gave us the opportunity to develop a conversation with government and with all of the people in the United States about the loss of valuable medicinal plants. It gave us something concrete to show that it is an internationally recognized problem. Without this, a lot of people who are connected to the medicinal herb industry will rationalize. They will say, well, you know, sure there is a problem with it on the fringes of its habitat, but it is locally abundant right down in the Ohio River Valley or in Kentucky or West Virginia. They will try to pull the wool over everyone's eyes and say this really is not a problem. The reality is that this is a real challenge, and we need to be farsighted in order to not follow in the tracks of, for example, the fisheries out west. Seven years ago, environmentalists were saying, "The Coho salmon is having a real problem. We are losing breeding grounds, and overharvest of timber is affecting the temperature of the water. The populations are obviously going down every year. Let's stop." Nobody stopped because they were rationalizing, "We always have had salmon, and we always will have salmon." Now it is being considered for the federally endangered species list—seven years later. So, we need to be a little bit more forward-seeking.

In a way, however, CITES did not function the way it was supposed to for goldenseal because there is very little export of goldenseal, which is what CITES regulates. CITES also limits export of the live plant and the whole dry roots, but it doesn't limit export of the ground goldenseal power or value-added products which are made out of goldenseal. It is really easy to get around the regulations on that side. On the side of whole plants, well, why would whole plants be exported? Only for cultivation purposes. There would be a possibility that those whole plants could be seeded into forests in England or in the Mediterranean or other appropriate areas in Europe or in the East. That could go a long way to save the plant. So, I don't really think that CITES worked very well for goldenseal in those respects.

I think a CITES listing would work very well for protecting black cohosh because there is a great deal of export of wild dug black cohosh roots from the American forests to Germany at this point.[2] It would bring the industry back home again, for one thing, to limit export of black cohosh. There is a big difference between black cohosh and goldenseal because huge amounts of black cohosh are exported and only very small amounts of goldenseal are exported, so CITES would have a much more significant effect on black cohosh than goldenseal.

MONOPOPULARIZING AND THE LIMITED MAINSTREAM MATERIA MEDICA

I think that the idea of monopopularizing comes out of a "this-for-that" kind of consciousness, which is really prevalent in the modern context. Generally speaking, people are taking a complicated study—herbalism—which relates to the use of potentially thousands of species of plants worldwide and putting it in popularized terms. The "top ten" plants is a common way to work with that.

I don't really have any problem with this phenomenon [of monopopularizing] because I think it is a beginning for people. It will prick their interest, and they can delve further into it. It also has some positive attributes, one of which is that then these plants tend to be the plants that are widely cultivated. Wild American ginseng, for instance, was at one time very rare, and without the large-scale cultivation of ginseng, it is obvious that the populations would have further deteriorated.

Another thing about the limited mainstream materia medica is that all you really need is a few good herbs. If you know a few herbs very well, you can address most minor ailments that would be the kinds of things that people usually self-treat. Although I wouldn't necessarily pick those herbs [laughs].

I feel a little sense of secrecy in terms of educating the public about the use of plants that I know are useful and not very common. I think it would be best for people to learn about herbalism bioregionally and get a feel for how they can integrate herbs into their lives without being part of the volume selling, health food store industry. It is easy enough to grow a few things outside your door which will address most of the health problems that you have. Calendula is an example, or there are some plants which are very common in the wilds like jewelweed which is an annual that grows right next to the poison ivy that it treats. So, I think that coming to understand the easily cultivated, very common plants, with which you can have a direct relationship, is a great direction and a solution to the popularizing of rare plants for medicinal use on the commercial scale.

PROACTIVE APPROACH OF UpS

UpS is a proactive organization. We are interested in educating the public and encouraging the manufacturing community to take a plant's at-risk status into consideration. We do this through the at-risk list and shelf talkers in the health food stores that say these are the plants that you really should cultivate if you are to use them. The at-risk plants really shouldn't be used out of the wilds anymore because there are less and less of them available. We are remaining really positive and proactive about the things we do, however. There would never be anything like a moratorium or a boycott. Those kinds of words do not enter our executive discussions. We want to stay in good communication with everyone who is affected by plants. We feel that by being a center for information about plant studies and about the status of the sustainability of various wild medicinal plants, we can do a lot more good than by trying to force the issue on people who are ignoring us [laughs]. I think that there is a good tendency for people to come over to our camp for that reason, and I think that we make a much stronger statement by inviting the discussion than we do by acting like we know it all and telling people what to do. I have had a few personal lessons in that direction that have shown me that my image of myself on a white horse battling the industrial dragon that is trying to consume the green world is something I can dispense with.

We are trying to evolve. Loving and respecting all human life and all beings is connected very strongly with loving plant life and all the green people. There is no "us" and "them." Basically, we are all creatures of the Earth and we are all sustained by the oxygen that the plants give off.

Chapter 9

Ryan Drum

Ryan Drum, PhD, MH, has been a professional wildcrafter, herbal educator, and practicing medical herbalist for over twenty years following a successful career as an academic research scientist in the field of cell biology.

Beginning in 1972, he studied herbal medicine with Ella Birzneck at Dominion Herbal College for twelve years; he has taught at the college's summer seminars for over twenty years. He was the clinic supervisor at the New Mexico Herb Center in Albuquerque (1998). He has been an adjunct faculty instructor in the Botanical Medicine Department at John Bastyr University since 1985.

Ryan believes passionately in true patient autonomy, in the complete freedom to choose one's caregiver(s), no matter what their credentials, and, that pleasure is the driving force of the universe.

Ryan and I spoke at the Green Nations Gathering, September 11, 1999, in Frost Valley, New York.

HERBALISM AND ITS RENEWED POPULARITY

We think of herbalism as a retrophenomenon in that it is, basically, a reaction against the industrialization of medicine. To have industrial medicine work, you need people to be industrialized. They need to be as similar as possible so that standardized patients are available for standardized medicines. In pharmaceutical production it needs to be that way. Pharmaceuticals are no longer compounded one little vial at a time. They were, if not originally certainly for the last four hundred or five hundred years of pharmacy. Before, let's say, electricity was common, the druggists were also called the chemists. I lived in the north of England in Yorkshire. A couple of these old pharmacists were still called the chemists. They were compounders. They would compound a medicine based on what a doctor or practitioner had recommended plus their own knowledge of what would be good for a particular diagnosis. That was individual prescription medicine basically. It satisfied even if it didn't cure. I think that what we lack most of all in our culture now is individual attention, individual interaction, individual opportunity to take charge.

Herbs as plants, many of them wild plants, also give a taste or a touch of the wild. We are kept from the wild if not by our cars, our buildings, our jobs, our spouses, our laws closing the lands off to us, then we are kept from them because most of our food comes from factories. They are called farms, but they are factories from my perspective.

The interest in herbs, I think, is an attempt to become self-reliant and in charge of one's own medicine. The sad thing is that those medicines are usually not free for most people because they don't know how or where or when to harvest them in spite of a lot of books. Most people don't read so much anymore, so books are unfortunately wasted on them. The market phenomena then of merchandising and selling herbs capitalizes on a fundamental need which is antimarket, in that the herbs are available—many of them just for the taking in waste and wild places everywhere, the places that can't be farmed. The lack of knowledge about how to get them and harvest them at the right time, the right place, and use them leads to market dependency.

Market dependency is what gives rise to lots of herbs and herbal preparations being available and few, if any, practitioners are available on a neighborhood or small-town or district basis to both educate and to administer those herbs. It is just like doctors in contemporary medicine, MDs, who are dependent upon manufacturers' representatives and the little flyer about how to use a medication for their basic information. Then people who would like to self-medicate are dependent upon just what is written on the label, or on laypeople who are the real practicing herbalists in America, who are not clinicians like myself and who still probably number less than one hundred.

CLINICAL AND LAY HERBALISTS

I refer to myself as a clinical herbalist. I refer to myself as someone who has a very formal education in chemistry, and in cell biology and formal botany. I taught those courses at the university level for a decade. I have a very highly trained, technical background. That was my root stock. On top of that I grafted twelve years of herbal studies with an experienced practitioner who had a large practice in Canada. She was one of the outstanding folk healers of Canada. I already had a very strong technical background and educational framework in which to put herbalism. I wasn't just starting out learning herbs. I already knew the plants, their cells, their lives. I wrote a dissertation on plant ecology. Not that I am great, but what that meant is that I had a very thorough education so that when I then started looking at the biomed use of herbs in practice, I already had almost all the same stuff that you learn in med school with the possible exception of making rounds. I did something similar with my teacher when she had residential care. There would be people sicker than I have ever seen anywhere else in my life. She would be treating them, primarily, herbally. Then, when I began my practice and I would see people, I would treat them as a medical herbalist. Later when I worked in a clinic in Albuquerque, I was basically a supervising herbalist the first time, and the second time I was the clinical herbalist as well as the clinic supervisor. I had student clinicians, those who wanted to be medical herbalists, sitting in and watching and listening [to] how I did intakes, how I diagnosed, how I prescribed, how I basically practiced herbal medicine. I then refer to myself as one of a few people in the United States who had not the exact experience of medical school but sufficient experience to be thoughtful, and cautious at the same time, being confident where I think I know. The unfortunate thing is that this is a very laborious process. I am over sixty years old now and I am just getting started. I feel, in my own mind like I have been in med school for forty years. The end is never going to happen. I will end first. It is continual self-education.

The lay herbalist many times is in the position of having to generate money based on herbal work or hasn't had the time or money to get a very technical education to understand what [he or she is] asking the herbs to do in terms of human biology. The lay practitioner is frequently someone who is working in a health food store, in a natural food store, in a co-op, in the herb section. Or, [he or she] may be working even in a large pharmacy or a so-called drugstore in the herb section. They are basically there to move product and to answer customers' questions. Those questions are the same or similar questions [as those] I am asked in clinical practice. What they have is basically a spontaneous or de facto herb clinic. They are seeing patients, diagnosing and prescribing, with little or no training, but with a lot of, usually, I feel, heart-felt positive intent. They are seeing most of the herbal medicine patients.

EDUCATION IN THE MARKETPLACE

In the meantime, the money, which is available for the purchase of herbs and herbal products, is too tempting for large merchandisers to say, "we can't sell this because people don't know how to use it correctly yet. We need to sponsor education clinics. We need to pay for herb teachers to go out in the neighborhoods, to basically do information missionary work about the herbs. Maybe they will buy our products; maybe they will buy somebody else's products." I don't know if it is greed or just a lack of thoughtfulness on the part of the marketplace, but I feel that inadequate education is going to lead to a relatively short-bloom period for herbalism because of poor-quality product, poor-quality diagnosing and prescribing without a medical diagnosis, as well as legal restraints against listing protocols on the packages of over-the-counter herbal medications. We need to provide the maximum amount of information along with a hotline to one or more computer programs such as HerbNET, where someone can ask a question, like, "I have been taking this tea for a week and all I am getting is diarrhea. I have been taking this tea for a week and I am not getting any better. Or I am getting better." There is data exchange as well as data withdrawal so that people are not self-medicating in isolation or in false hope. I think that people should do the best for themselves based on the best information available rather than the cheapest product or fad medicine. Fad medicine is the curse of the uninformed culture.

NECESSITY OF A THOROUGH EDUCATION
FOR PRACTITIONERS

I don't know if it is just scientific background [that is necessary for herbal practitioners]. I think it is the humility that comes with a thorough education. There is enthusiasm fueled, in part, by ignorance. I think most practitioners who have been in it for sometime, even a few years, but certainly five, ten, or twenty years, will generally admit that the human body and its presentations are the most complex things in the universe. Even though we are part of that presentation, it still remains mostly mysterious. The things that seems to work, work largely on faith and hope and trust as opposed to several studies with thousands of patients showing that certain dosages of a particular herb containing certain constituents will cure or relieve symptoms. We don't know that. Every patient is an experiment in that sense rather than a guarantee. I think unfortunately in our culture we are led to believe that medicine should be guaranteed. I think herbs are always a gamble and that is part of their appeal and risk, as much as the low cost.

I don't think everyone needs to go and take four years of chemistry and five to seven years of cell biology, but a little bit would help. There are several schools that are trying to get enough of a technical education that herbal medicine could be practiced with some basic science understanding. If you are a little unclear as to how solar radiation presents and/or operates in the visible spectrum, in the invisible spectrum, then there is a lot more mystery than is necessary. Anything that will enhance your knowledge base for dealing with a particular presenting pathology will improve the quality of how you deal with that. It is also important to explain what is going on to a person you are trying to help. Many times I find in clinic, it is not just dispensing herbs that I do. I do two things: One, I provide human energy to people who, if they had enough, would be healing themselves. They would be getting more than enough sleep, eating good food, drinking plenty of pure clean water, and being in loving, satisfying relationships only. That tends not to be the case. The second thing is that I am giving them an education about the pains and discomforts that they feel and the probable origins and developmental sequences that led to those pains. If they have little or no understanding about the origins of their pathologies or pathology, then they are likely to repeat it one, or two, to dismiss some of my suggestions. So, I spend my time in giving energy and education. It is almost as though the herbs were merely a charade, or a vector for life and lifestyle advice. I don't think of myself as living perfectly or being in perfect health. I am continually self-medicating that way as well.

In many medical schools, and naturopathic schools as well, the educational process itself becomes painful. The pain I am referring to is recognizing the need of people in distress, and recognizing one's own limitations of being able to deal effectively with that. That is taught in med school, but the humility of being bludgeoned for not knowing is a false thing, or an erroneous thing. We are all ignorant of something. We are all very ignorant at one time. Every doctor, every genius, every wizard was dumb, maybe not stupid, but dumb at one time. We must remember that with humility so to be loving and compassionate in education, but we must not gloss over the fact that everyone dies. Some people die by their own hands, some by the hand of a practitioner, some people will die with no apparent reason at all. Those deaths are real. They involve real people. It is not just about having miracle cures from giving comfrey or nettles or something and having everyone be happy ever after.

If you are in a store situation, as an unintended lay practitioner, many times you may never see those people again. You may have to quit your job or leave because you want to earn more than seven dollars an hour, or they may die or get run over. You have no idea. There can be a false sense of having helped these folks because of lack of contact.

The initiation for medical herbalists that I envision may be a little bit shortened from what I did for twenty-plus years. That is why I would like to see herbalism have semiformal schools where you have access to patient rounds. Medical students would spend time at herbal schools taking herbal classes, not from the med school perspective but from the herb school perspective. Herbal and medicinal schools would trade students, and herb students would be required to attend autopsies, to do a cadaver dissection, and basically to handle and see the reality of human frailty in terms of tissue.

One of the most distressing things to me in the teaching of anatomy and physiology just by pictures at herb college was that there was a false sense of lightness on the part of the students. In 1977, we started doing animal dissection at Dominion Herbal College. I did a dissection of a sheep, starting with a live sheep. It made noise and was lovable and cute. Then it was killed and half skinned so they could see what the animal looked like without its clothes and also with the fur still on one side. We proceeded to go through gross anatomy, muscles, bones, organs, how things were placed, and design failures. Students also got to feel warm flesh. It wasn't vivisection, which would be, ummm, possibly contraindicated, but students realized that they were going to be dealing with the same thing with a little higher IQ, a biped instead of a quadruped, but real live tissue. The same organs, the same, oh my gosh . . . it was an amazing experience. People have written to me and come up to me decades later and said it was one of the most wonderful things that ever happened to them. They had never seen inside a freshly killed animal before. They had no idea that was what muscles look like, or organs that were still basically twitching, or peristalsis still occurring. Not that I thought what I was doing was great. It was just the bare minimum. Tissue reality. To practice herbalism without tissue sensitivity might be effective, but it is not patient positive.

The other thing to do would be to have if not a resident program, but some type of obligatory real apprenticeship or preceptorship, where you need to sit in for at least a month with a practitioner and pay for that if necessary. Having had preceptor students, both herbal students and naturopathic students, it can be exhausting explaining things to them or just having another person in the intake room. They can get experience with somebody who is seeing patients regularly. I feel that it is foolish to call them clients. If we are practicing medicine, which we are, then they are our patients. We are diagnosing their conditions, if we can, and we are prescribing herbs.

CLINICIAN-PATIENT RELATIONSHIP

I feel that the patient needs to work at least as hard as the doctor, if not harder. If I have a noncompliant patient, then I usually withdraw. If I am expected to devote one hundred percent of my intellect and my understanding to the care of a patient then I am expecting [him or her] to participate one hundred percent. It has happened—if not a lot, certainly frequently—that people have said to me, "Ryan, I can't devote my life to getting better." I have had to say, "Then I am sorry, I can't devote my expertise to your partial participation in this. Your life is at stake. I wish you would pay attention." Some of those people have died, sometimes relatively soon after seeing me, from if not preventable certainly ameliorative conditions. One never knows. After they are dead you can say anything about it. So, I don't think that I accept responsibility for anything other than being as accurate and thorough as possible in what I say. It is the [individual] that actually does the healing. I am merely a vector for information and energy and glad to help out.

In terms of authority figures, it is the curse of the knower to try to avoid, carefully, a dominance hierarchy situation where those that don't know must submit to those that do know. I feel that is psychologically very destructive to doctors because they begin to believe their own assumed role, that they really are an authority rather than knowledgeable and privileged. A submissive patient is not necessarily a compliant one. If at all possible, I like to have a sense of equals. My preferred practice is not the clinic. My preferred practice is a neighborhood or village or urban community where I tend to see the same or similar people regularly over the years. I have some idea of the etiology of their diseases, how they develop, or their pathologies because I live in the community too. I am experiencing similar stressors. I am experiencing similar disappointments or noises or sewage failures, not myself as a person, but as a resident.

I would like to offer my services without money. In the community where I have lived for twenty-five years that has tended to be the way. There was a time when I charged a fairly substantial amount, and I felt that it interfered with the quality of my practice because there was always the dollar sign up in the corner. There was a little double slash *S* sign up there. That was what these people represented even though I didn't necessarily want that to be. That was what was happening. I found if I mentioned that I would like something equivalent in exchange, once that became known in the community, then that happened many times without mention or with a few details worked out. I was not offering free medicine but equivalent medicine. Everybody has something to give. Even if it is a pack of lies, everybody has something to give. I felt that whatever they gave would be their best, and I like to assume that. Over the years that was what tended to happen. Instead of me getting a check or a pile of dollars, they could say, "Well, I paid him, little bastard. He didn't fix me." They gave something from their lives, some energy, something they had made or done, fish that they had caught, food that they had grown, a massage, a painting. There are always things that need to be done that, upon spending time helping someone get through a bio med difficulty, I don't have time to do for myself. I was happy to trade my expertise for theirs.

Eliminating the money exchange meant that I got to see them usually much earlier in the development of a pathology instead of by the time it is too late or it really needs serious intervention by someone who is into what I would call more intense medicine than I am. I try not to pretend that I know how to deal with acute medical emergencies that require either special equipment or very special knowledge. I do some first-aid work. I have advanced first-aid training. I think, mostly, I deal with subacute or acute-slash-chronic conditions. We have some idea what is going on. Occasionally there will be chain saw accidents or sores that won't heal, acute things like blood in the urine or incredible pains, some heart problems, but because where I live requires a good physical condition to begin with, I tend not to see in community practice what is seen in clinical practice. To be in the community, you need to be healthy to begin with. I tend to see long-term self-abuse things from lifestyle. I also have some idea of if they will be compliant or not. That is a part of my participation as a practitioner.

I also harvest almost all of the herbs that I use in practice. What that means then is that I have a great faith in their quality in terms of purity, medical efficacy, and previous successes. A knowledge bank gets built up, not just of using a particular herb, but using the herb that I myself have harvested. If it has been stored for too long, and I assume that its efficacy has failed, then I use it in a bath or a soap or as compost or mulch in the garden. I find that a very rewarding, personally, type of practice.

Wherever I go, there seems to be an instant clinic set up. There is a line of people waiting to talk to me in part because, especially if I am at a conference or in a teaching situation, I don't charge at all. I feel that I have been gifted with a good brain and a healthy life, and it is mine to share, not mine to hoard. If I were to go sit off and say, "Oh well, I am smart and you are dumb. Too bad. I will choke on my own knowledge." I feel that sharing is obligatory for good health. I feel somewhat sad that the contemporary major medical establishment seems to feel that hoarding or protecting information is a smart and healthy thing to do. I think it is a consequence of the so-called capitalistic, or what is really the market, monopolistic economic system. If there were less sequestering of resources by a few, then the rest of us could probably live in a little more relaxed and healthful way.

USEFULNESS OF SCIENCE AND FOLKLORE

As a scientist severely trained and having worked as a professional chemist and cell biologist, I would say that science is extremely limited. It attaches absolutism to guess work. With the possible exception of gravity on the surface of the Earth, everything is an approximation.... As far as I know, all scientific theories start out with "if we assume." Those three words mean: "I don't know for sure." Everything else that follows is guess work. Science works only so long as it is empirically provable, that you can either repeat an alleged observation or the next observation seems to approximate the one you saw before. Since nothing is ever the same, everything is different. Every reaction that claims to be the same is approximately the same; it is never exactly the same. Fortunately. Heisenberg uncertainty principle plus the Pauli exclusion principle.[1] I think those are the kinds of things that are important for herbalists to understand, the limits of science and what science is trying to do.

All folk medicine is based on two principles, from my perspective. One is trust in constructive ignorance; that is, when you don't know but you have an idea. I have no idea where that idea came from other than it is suddenly there. Use dandelions to cure hangnails. You don't know how come. But you have been using dandelions for thirty years and your hangnails disappeared. You forgot about that. Someone comes to you with terrible hangnails and dandelion is the only thing you can think of. That is what I call instinctive knowledge. We all have that. Very few animals die from ingestion of poisonous plants if they are allowed to browse on their own. Even ones that have no previous experience with those plants will not eat them. They are detecting the potential lethality of ingesting a particular plant. This doesn't always work as we know, but in the wild, very few wild animals die from poisoning from plant material, though.

The other major knowledge tract is knowledge gained by empirical experience. You eat belladonna, then you are going to have a little bit of heart palpitations. Your eyes are going to dilate. You may look sexy, but you are going to feel sickly. Same way if you eat a certain mushroom, you may be dead. If you eat half of it, you may have the pyschotropic experience of your lifetime. Funny, how did they ever figure out that your urine is going to be even more exciting than the mushroom ever was to the next person and to the next person after them, maybe even more exciting 'til seven times passage? Now, it is very difficult to have an intuitive sense about drinking the urine of someone, who is having a psychotic episode, as being a positive experience. There has got to be some experimentation there. In medicine we are presented with an obligatory set of semi-independent variables as well as buckets of unknowns. The security needs of the scientific mind in its purist form are such that it is intolerable. There is the clash between folk healing and/or folk herbalism and scientific herbalism. Scientific herbalism does not feel comfortable with "if," "maybe," "sort of," "perhaps." And yet if you try to apply scientific medicine to patients, everybody responds slightly differently. And so, to be practicing really good medicine means practicing or using medicine person to person rather than the same medicine for all persons. Each person has to be dealt with as a person. The folk healer or folk herbalist, many times because of time availability, has opportunity to . . . spend time with a person, but also to encounter [him or her] as a person presenting a pathology rather than a pathology attached to a person. I don't think there is a common ground. I think as long as there are two polar extremes, there will be no common ground. One is the destruction of the other.

FORGING A RELATIONSHIP WITH THE ALLOPATHIC COMMUNITY

If you have a license to practice medicine, you have a license to kill. Very few doctors are prosecuted for medical homicide and yet somewhere around two hundred thousand people die each year as a consequence of various types of medical interventions, through prescription medicines and the tens of thousands of people that die from too high an air pressure in the operating room where they cannot exhale.[2] The apparatus is forcing too much air into them and nobody is paying attention to the machinery. All of those deaths are ignored. Just like childbirth deaths due to separating the base of the brain from the spine because the obstetrician too hurriedly just pulls the head out and separates that delicate tissue. Those are not called homicide, but I believe they are. A medical license carries with it access to resources of income and also a license to kill. As long as practitioners are given a license to kill, and access to material resources that the rest of us do not have, then there will [be] inequities. I believe it is the economic system or material system of culture that generates the medicine that it gets. The medicine system we have is a consequence of the economic system. One of the reasons that it is so amazing is that it is mostly battlefield medicine. It is crisis medicine. It is heavy intervention. No other animal voluntarily submits to surgery other than the human being, and yet most of us just seem to do it with little or no thought. That is basic denial of self-integrity, to allow yourself to be sliced open. The message to our body is nobody cares: "This person will just let me be completely cut open by strangers. Why should I bother to heal? Why should I bother to be well? This is how valued I am. I can just be cut open." That is the basic message—that you are expendable.

I don't want to bridge to people who think that is fine, that it is an okay way to practice medicine or to set up a culture where impact trauma through automobile collisions is an expected event. I see the whole milieu, the terrain in which allopathic medicine develops and is practiced, as pathological. Civilization from my perspective has been found inadequate and wanting. I don't think we need it. It is a "doomed to fail" use of resources, including human capacity. Most of the patients that I see are not sick from overuse, except for maybe carpal tunnel syndrome; their bodies are wasting away from the betrayal of design intent. They are not walking; they are not running; they are not climbing, except as a charade of exercise. Behavior with no intent. Whole body masturbation. It is bizarre. We have made ourselves unnecessary through our very cleverness. Only the stupid will survive because they don't have enough smarts to be anxious about it.

OFFERING HERBS WHEN ANTIBIOTICS ARE NOT APPROPRIATE

Well, if a patient doesn't need antibiotics (and most of those people, I suspect, were presenting with viral conditions, which are not amenable to antibiotic herbal, or otherwise), then the doctor is being dishonest [in prescribing an herbal alternative], but he [or she] is kidding them along. Everybody is participating in the charade. The emperor has no bacterial infection, but we will give him antibiotics anyhow, herbal or otherwise. I believe it is a misuse of herbs. It is a misuse of position. Instead of saying, "Well, there is really nothing that I can do. I won't charge you because I am ignorant of what can actually be done for your condition." Rather than just giving them something, why not say, "Sorry this one is free." I think charging people for your ignorance is terribly cruel. The other thing is that because we are no longer a community-oriented culture, but rather a consumer or materialistic culture, we don't take care of each other. So, doctors are being asked—it is demanded of them—to do all the things that used to happen in just human interaction: "Hello. How are you?" "I am horrible." "What is wrong with you? Let me fix you something. Can I come over and help you?" Instead you go to the ER [emergency room] when basically your heart palpitations are caused by loneliness, and you self-medicate by eating a pound of salty food or a pound of sharp cheddar cheese or two bottles of very expensive wine. That is lack of human community. I think doctors are being asked, and also pretending, to give more than any one human being can at any one time. Most of the conditions that come into clinic are lifestyle pathologies. Those lifestyles are dictated by economic reality. Economic reality is dictated by greed of the few at the expense of the many. That is real encouraging, isn't it? The circle continues round again.

Chapter 10

Daniel Gagnon

Daniel Gagnon has been a practicing herbalist since 1976. Canadian born, he relocated to Santa Fe, New Mexico, in 1979 where he studied medical herbalism, pharmacology, and related subjects. Over the years, he has become an accomplished speaker and educator, giving classes and workshops across the country and abroad.

Daniel is also an herbal businessperson. In 1982, he purchased a retail shop called Herbs, Etc. from herbalist Michael Moore, and began formulating and manufacturing liquid herbal extracts under the same name. Today, the Herbs, Etc. retail shop is thriving, and the Herbs, Etc. manufacturing facility has grown to over 7,000 square feet, supplying a line of 132 single and formula extracts to retail natural food stores in the United States, Canada, Mexico, and Holland. Daniel is also a senior professor of materia medica at the North American College of Botanical Medicine in Albuquerque, New Mexico.

I spoke with Daniel over the phone on October 6, 1999.

HOLISTIC HEALTH

To me, holistic herbalism basically addresses more than the symptom that has manifested. That means that the herbalist has to look beyond what the client is presenting. Oftentimes, there are a lot of other issues. One of the classes I give is called Elements of Good Health. I go through certain aspects of health including things like rest, exercise, emotional support, having a base of friends and people to talk to, having goals, having a spiritual connection of some kind. I want to address all of that as I am addressing health. That is holistic.

One of the things I have found over the years is that most problems are fairly simple. The medical field makes it seem like it is getting more and more complicated. In reality, if you step back, it is more and more simple. A health problem is simply a lack of balance in some area. Yes, there is some genetic predisposition to certain diseases, but if that were strictly the case, everyone in a family that had a parent with cancer would get that same cancer. It doesn't necessarily manifest, however.

What are the differences between allopathic and holistic healing? That is where I come back to the place where we have to make sure we are addressing those basics. They are not only on the physical realm, but also on the emotional and spiritual realm. I don't know how to separate that. I know it sounds like hocus-pocus, but think about simple stuff—for example, exercise. If you go for a half-hour walk five times a week, you start activating your blood flow. Your lymphatic system starts draining. Your lymphatic system can only drain through the use or manipulation of muscle or, for example, through massage or exercise. When you activate that muscle, it helps to drain the lymphatic system. When you walk, you also get the fresh air. You get some sunshine. You get tonification of the abdominal muscles. That will help with the bowel. It is incredibly simple. A half-hour walk, but the ramifications of that walk five times a week are great.

What I usually say to people is look at those ten elements of good health. Then choose the one where you are the weakest. Start working on one of those. Some people, for example, may eat an excellent diet, exercise, get enough sleep, but guess what? They don't have any friends. They don't have people to share their life with. They are isolated. Well, we know what happens to preemies that are put into those little incubators. They don't do well. They don't gain weight. Health wise, they tend to fail. If you hold them, simply holding them—they start gaining weight.[1] They start getting well. So, why would we ignore that fact? To me, that is part of health.

Another aspect that I think I need to address is the largest part of that holistic pie of healing. When you look at those different categories, the most important by far is *choice*. Forty percent of my holistic pie comes from *choice,* the daily choices we make. Am I going to eat at McDonald's today, or am I going to get a salad or some whole grains, even a potpie or something that is healthy? Or, am I going to have a hamburger and fries and a soda? That little choice, that tiny choice: am I going to exercise today or not? Those choices add up.

The concepts of holistic healing are extremely simple. That doesn't make them any easier. It is so simple.... When I say to people, you need to exercise five days a week, they nod their heads. They go, yeah. Do you do it? No. Are you going to do it? I am going to think about it. Then you have to somehow motivate them; you have to show them why it is so important; you have to show them what it does to the heart, the lungs and the blood circulation, and the lymphatic system and digestion, et cetera. Some of them will start to go, "Oh, I didn't realize." Exercise is just exercise to them. It doesn't have all of these other implications.

What I usually end up doing, especially in my classes with the herbalists that I teach, is I give them an idea of how you can help people to set goals. I use a system based on a week. The person does something six times a week, and then [he or she] get[s] a reward. If you do it six days a week, like you do your exercise six days a week out of seven, then you have reward. I will give you a simple example from my own life. My goal was to floss. I never flossed before. I couldn't get it into my system. I would do it twice or three times a week but not on a regular basis. The goal was to floss six days a week with a one-day rest anywhere in there. If I flossed six days a week then I could buy a CD [compact disc]. After three months, I didn't need to buy a CD anymore. It was a habit. That is what we need to instill in people. These are simple things. Here is your reward. After a while, it becomes a way of life.

THE PROCESS RATHER THAN THE PILL

What we are talking about in some ways is more natural medicine than herbalism. Herbalism is the use of herbs for health problems. My thing is that if we do it that way, we are just like a medical doctor. Here is the problem; here is the herb. We haven't solved anything.

I think oftentimes that herbs are probably a placebo. I often think the same thing about a drug. Just the fact that a person is, every morning, taking, dividing [his or her] pills and taking those three pills a day. Maybe the pill has no action; it is the action of taking the pill. There is a movement that goes even further. It says that the therapeutic aspect of it is the water that you drink with it. There is a book out there called *Your Body's Many Cries for Water* (Batmanghelidg,

1995). It says that most of us are in a state of semidehydration. Oftentimes it is the water that you are putting into your body that has the curative effect. The doctor gives you six pills to take, you drink six half-glasses of water. That is the therapy. It is wild when you think about it. I think that a lot of our herbs probably—I could even guess that a lot of our herbs do not work. And neither do the drugs. But the doctor giving it with a certain confidence or the herbalist giving it with confidence . . . we know that has a lot of ramifications.

You know, I believe in herbs. I have seen them work, but I also believe that maybe it is my belief in herbs that makes them work. In the same way that taking the pharmaceuticals is self-affirming.

MORE ON HOLISTIC HEALTH

I am saying that the other elements of health are as important as herbs. For an herbalist to start crowing that if only the world would turn to herbs, everything would be okay—I don't believe that. I believe that I need my men's group to help support me when I am going through hard times, that people at work . . . I want to make more than just a work connection. I want to make a human connection. I want to have contact with a higher being. That helps me when times are really tough. It doesn't matter how many herbs I take when times are really tough. They will support me and they will help me through it, but it is only one thing. You need to talk about issues with somebody else; you need that support. You have to take a deep breath, go out in the mountains, and think everything is okay. It is not the end of my life.

I think most herbalists appreciate that. I think most herbalists will talk about other aspects of healing, body work or meditation, or exercise; they will usually include other elements.

[Doctors] usually will not do that. [They] will either say surgery, chemotherapy, radiation, whatever they have learned, but they usually don't branch out much because they tend to specialize. We herbalists tend to be more generalists. I think there is a place for both. I am not against orthodox medicine in any way. I use doctors a lot. People come to my store, or they come to me, and they say I don't know what I have got. I say to them, first go and find out. Once you have what I call a label, then we can proceed. It is a protection for the patient. What if he [or she] has something that needs to be taken care of surgically? That is possible. Herbs won't help, but there are a lot of times when herbs would do very well.

IS THE AMERICAN PUBLIC HEALTHIER TODAY?

I go in my local grocery store in Santa Fe, a regular grocery store, and I would say that now one-third of the offerings in the fruit and vegetable aisle is organic. Wow! Amazing! This is cool. I can just run to the grocery store and get some organic spinach and organic something else and run back home and cook it. Conversely, there is all of this fat-free, sugar-free, salt-free, everything-free food.

The same thing with over-the-counter drugs and the herbs. If I have a prostate problem, all I have to do is pop a pill. Or, I can become more conscious and say okay, I am not exercising enough. I should drink more water. I need more fatty acids in my diet so I am going to eat more salmon. Then I get healthy.

There is both. There are some people who are switching from a drug to an herb . . . it has the advantage that it has less side effects, but how much progress is that? Conversely, it is expensive. So, it is a little bit of both.

HELPING THE MAINSTREAM ACCESS HERBAL INFORMATION

Herbalists are doing a lot of teaching. There are more herb books. The information oftentimes is available. Sometimes it confuses people because they read about macrobiotics, vegetarians, the raw fooders. There are so many diets. There is the Pritikin, which is real rigid, no fat; then you go to Atkins, which is basically a lot of protein and no carbohydrates. It makes for a confusing world out there. Herbalists can help direct people in some ways. We can also educate them that it is more than just isolated health problems; it is lifestyle, lifestyle, lifestyle.

MEDICAL SCIENCE AND ITS INTERFACE WITH HERBALISM

I like the fact that the medical community is open to herbal healing, that they want to learn. I hate the fact that they all want standardized herbs. They feel that there is only one active constituent. The whole plant worked for five hundred years. That is how we know about it in the first place. It never was standardized, and it worked. Oh, but we need to know what is the active ingredient. They isolate for one constituent even though it may not be the "active" ingredient. Like St. John's wort with hypericin, which we know is not the active ingredient, or feverfew with parthenolides, which are also now no longer the "active" ingredient. We have no clue. The physician, the researcher, I think, has to look more profoundly than that. We need to keep pushing as herbalists to ask for a more profound change. Stop looking at the cells. Start looking at the person.

SCIENCE VERSUS FOLKLORE IN EVALUATING HERBS

I think herbalists are interested in scientific studies of herbs. I think we do want an "in" to some studies, but . . . also, if you look at studies, they are very limited.

The traditional knowledge of herbalists contains a lot of information which medical science is hesitant to pick up. We keep saying this over and over. We keep saying that people have problems because they do not drink enough water; they don't have enough fiber. How many years did we talk about not enough fiber in people's diets? Boy, is it recognized now? Oh, well, now we have all the scientific studies to back it up. But it had to come from somewhere. So, we need both. I am not for just one or the other, science or folklore, but let's not get stuck in either model. Let's keep on searching. Let's try to find out, but let's look at whole plants. We like whole foods; let's look at whole life.

HOLISTIC VERSUS ALLOPATHIC HERBALISM

People ask, "Can I replace Lasix with dandelion?" Well, yeah, you can. Then they say, "Okay, thank you. That was all I needed to know." Well, it is more than that. Why is that person having kidney problems in the first place? That is not my problem. I just want to give him something to get rid of the water, says the physician. Herbalists go a little further. Most herbalists are looking a little further. We utilize plants because they have been a way to bring in people. We have less to worry about in terms of side effects, less about contraindications. Now herbal medicine is starting to go that way big time. The list of contraindications and side effects for each herb is getting pretty long. It is going to keep on getting longer, but overall, herbs are not that dangerous. You can do the same thing with vegetables or fruits. Probably there should be a warning on fruit. Or, if you have a goiter, you shouldn't eat any cabbage because it is goitergenic, but we don't do that.

MAINSTREAM ACCESS TO HOLISTIC HEALTH

I think books are a real important way for the mainstream to learn about herbalism. For a female who is going through menopause, they should get Susun Weed's book, *Menopausal Years: The Wise Woman Way* (1992). It is a great book—a little radical for some people, but I say look at it as a sum of information. Take what is useful and throw the rest out, but in the process what that does when you read those books is that you expand your mind.

One person who has done a tremendous amount of work in the last five years is Andrew Weil. He is extraordinary because a lot of the ideas that I have been teaching for quite a while he put even more form to and devised a program. A lot of people are trying that program and are feeling better and incorporating positive health elements into their lives. One of the things that he does is a news "fast," where you don't listen to television or radio or read the newspaper. People's anxiety comes down. You are bombarded with all of this information about war and disaster and you feel for people, but you feel like you have no control, like there is nothing you can do. You feel out of control. Anxiety level goes up. When you wean yourself from all the news, and you come back to what is happening locally, then you can control it. In some ways, what [Weil] is prescribing is extremely simple. But it is very effective. When people want to learn about holistic healing, I direct them toward those books. I find books that I feel have a philosophy of life that is all encompassing and I suggest them.

The other thing I tell people is you are your own best doctor. You have to become a detective. Find what doesn't work for you. What makes your problem worse; what makes it better? Try to meditate on those. Try the different herbs; try the supplements. Find the program that works for you.

LICENSURE FOR HERBALISTS

I am all for licensure. There are some issues, however. For example, in New Mexico, we have a lot of what you call curanderos, who are basically the native healers. They don't have training. They don't know about anatomy and physiology. They talk about twists in the intestines. They talk about a lot of different things that we don't have a vocabulary for, but for them it makes sense. They know which herbs to use. They know how much to use. I do not want to prohibit these people from practicing.

What I would like to see is a standard exam where there is a degree of knowledge that is evaluated. You would need to have anatomy and physiology, some biochemistry, some pharmacy science, some pharmacognosy, botany. You should know one hundred fifty plants. With that exam, if you pass, you are able to call yourself a "clinical herbalist" or a "medical herbalist," or a title of some type that you put on your business card. It is like a registered dietician. It means something. It doesn't prohibit somebody else from calling himself or herself a nutritionist, but they cannot call themselves a registered dietitian. That is the way I would like to see licensure, where a degree of knowledge and experience is recognized for specific healing systems, like Western herbalism or Chinese herbalism or Ayurvedic herbalism.

It is fine if [other herbalists don't] want to be licensed, but they should not call themselves medical herbalists. They can call themselves "wise woman," if that happens to be, or what they identify themselves as. I am fine with that. It is like the registered dietician and the nutritionist.

People will gravitate to what feels good to them. Like, why do some people go to an acupuncturist or an Ayurvedic practitioner? Have you ever had acupuncture? I don't do so well with acupuncture. I passed out twice, so I am not real keen on going. Maybe it is too energetic for me. I do really well with homeopathy. That is my healing of choice. I have a naturopathic doctor who does only homeopathy. That and chiropractic works really well for me. I gravitate toward that. Another person will say, oh no, you have to go to an acupuncturist. Man, that works. For you, it works. We find what works the best for us.

EDUCATION FOR HERBALISTS

The most essential piece of herbal education is definitely the materia medica. That is the base where you learn about the herbs themselves. [Herbalists] would also need to understand pharmacy so that they know how to prepare medicines properly—also, anatomy and physiology so they know how the body works. Maybe some biochemistry so that they understand some of the reactions that the plants have inside the body. Also, diagnosis so that they know when to send a person to a doctor. Clinical practice is important so they know how to sit with a client, how to talk, how to have some relationship with the patient. Those types of things I think are critical to a program.

The advantage of learning in a school is that usually you have a curriculum. There is a certain amount of topics you are going to cover. [People] may not retain it all, but they know it exists. If they have questions, they can go back and research that aspect. What is missing is the day-to-day working with that plant. That is where the apprenticeship is stronger. You go through the steps of harvesting, grinding, extracting, whatever, many times. You sit with the practitioners and hear the questions they are asking, how they are asking the questions, and how they choose their herbs. Perhaps we should incorporate both styles of learning: classroom and apprenticeship.

I made a proposal to a well-known herb school to have the second- and third-year students work in my retail store. They would fill the jars of herbs. They would talk with people. They would mix formulas for people. We have behind our counter two hundred fifty bulk tinctures so that we can do special formulations for clients. We would give the students that practical aspect. People learn a lot when they "get the dirt under their nails."

In a retail store, students could interact with clients. One of the things you learn in the retail setting that you don't learn in the clinical setting is to think really quickly. You have got about three minutes to evaluate, roughly, what is happening and give a suggestion. It is a little bit symptomatic but you learn to think about the herbs quickly, to go through the list of thirty maybe that you have in your head and narrow it down. That helps you discriminate.

The other thing that happens in a retail setting is you get close contact with all of the herbs. You get to smell these herbs, and you get to play with them and put them in the jar. After a while, after replacing the chamomile in the jar ten or twenty times, you go, hey, that is a good batch. The batch before that wasn't so great. But you can't even tell how you got that. You just refilled that jar and refilled that jar. And finally you noticed it. That is what happens with the practical aspect. Or, you mix tinctures and the client goes home and comes back and says wow! This really works. Or, comes back and says, you know, my problem is still there. Then you have to delve deeper or you change the formulation or you find that you were on the wrong track.

ENDANGERED PLANTS, ORGANIC GROWING OF HERBS, AND DIVERSIFYING THE MAINSTREAM MATERIA MEDICA

Herbal medicine is exploding. The amount of black cohosh, goldenseal, and St. John's wort that is being used is astronomical. We have to grow those herbs if we want to have them for future generations. Most of what we use, why can't we grow it? That is what I have been encouraging people to do. We are starting now to get some cultivated goldenseal. All of our blue cohosh is certified organic. All of our St. John's wort and mullein, et cetera. People say, mullein? Why do you even bother with that? It is everywhere. Because we should get everything grown. The farmer makes sure that next year he is going to have a crop and the year after that. It is not always the same for wildcrafters. That is why I am a strong advocate for certified organic growing.

I think it is about education primarily. What I say to people who are using echinacea as a preventative, "That is astragalus's job; that is schizandra's job." All these different herbs. Goldenseal: don't use it as a preventative. Don't use it on the first day. Use it on day three, four, or five. It is specific for inflammation of mucous membranes. If you are at the beginning of a cold or a flu, you hardly have any inflammation. You have a lot of immunological stuff happening, but not much inflammation yet. Later on, goldenseal would be appropriate. Then if the cold is hanging on there are other plants. There is garlic; there is red root; there are so many other plants. What I try to teach and what I try to write about is how to choose the right plant. Right now we are just giving black cohosh to everybody. Well, sometimes dong quai would be more appropriate. Sometimes angelica root would be more appropriate. Blue cohosh perhaps. It depends on what is happening. The education—that is a tough job! We are making progress.

I love to see the farmer's market. That is also part of the movement. It used to be that you had only one type of corn. There are a lot of different types of corn, and they all taste different. There are a lot of hot peppers, or a lot of different apples. We were getting down to a Granny Smith, MacIntosh, and maybe another one. Now we are re-expanding. I love all of those different varieties. That is what we need to do with medicinal plants. There are many other choices. Especially also if we grow them commercially, certified organic, then we don't have to worry how much we consume. We are growing it.

Chapter 11

Leslie Gardner

Leslie Gardner is a master herbalist who had the unique opportunity to study at the university level with Sidney Yudin, a Russian-trained doctor of botanical medicine. After completing her degree, Leslie practiced as a clinical herbalist in Los Angeles and Berkeley, California. During that time, she created and marketed a line of herbal teas in the San Francisco Bay Area.

At present, Leslie is the coordinator of the Garden Apprentice Program at The California School of Herbal Studies. Her classes include horticulture, including all aspects of cultivation and propagation of medicinal herbs, magical herbalism, doctrine of signatures, as well as the lore and history of botanical names. Leslie also manages Emerald Valley Herbs and is the director of the Sonoma County Herb Exchange.

Leslie and I spoke over the phone on October 1, 1999, on the topics of herbal education and spirituality.

LEAVING CLINICAL PRACTICE

I don't practice anymore as a clinical herbalist except informally. What took me out of practice was what I used to call the umbilical cord syndrome. I tend to have a larger vision of what I wanted people to do when they came to me. I found that, by and large, most people just wanted to be told what to take at what time. I was more interested in them looking at the larger patterns that were developing in their lives and their health, in their mental health. It was a larger picture that I was interested in. It was very exciting to me to discover that.

My training in Chinese herbology, in Chinese traditional medicine, helped me to see some larger patterns that we don't see so much in the West: connections between organs, parts of the body, and certain symptoms that aren't really very apparent to us here in the West, with our Western diagnostics. I would be very excited to share that with ... [clients] so that they could take that into their lives and begin to see some patterns themselves and begin to work more proactively with their own health. In other words, I would say, well, you come to me for something. Maybe we have worked on that, and that is good. Now, would you like to take it a step further? Would you like to take a look at your life and the patterns that may be developing, that may develop over time as you get older because of what I have seen, because of what you have shown me, some of the pathology, some of the words, the season that things come up in? I have noticed that this may be a weak area for you over time. This might be something that you might want to think about, in a more macro sense. It was disappointing to me over and over again. There were a few select people who were very excited by that concept, but most people weren't.

I am getting back to the idea of spirituality because to me those ideas, those patterns at least start to dig into the spiritual. It is starting to look at something beyond the physical, beyond this quick fix, onto what is my life's meaning? What am I doing? What is my intention? Where are my resources?

WORKING IN DEEP RELATIONSHIP WITH PLANTS

I have always seen plants as allies. When I was in clinical practice, it was more about little pills and tinctures. It wasn't outdoors in the garden the way I am now teaching. I wasn't so much talking to the people about plant allies, although I did to some degree. Now I work directly with the gardens all the time. Now, it is an extremely integral part of what I do. It is an integral part of my classes as well.

I feel that oftentimes the plants are working through us and that we don't know that they are. But they always are anyway. That sounds very "New Agey," but I have been convinced. It is sometimes stuff that we just can't sense very easily. In our cultural model, we have come to believe that it is us that has to come up with the ideas. Oftentimes, I have been very humbled by the fact that, particularly in my work at the school and with students and my direct work with the plants, many times when I think it is me, it is really the plants that have come through me. I encourage the students to do that same thing.

James Green and I start out a class together in the beginning of the school year in which we introduce to the students the concept of communicating with the plants.[1] It is their first introduction to that concept. One of the things that I often say is that you can begin to trust your intuition. And begin to recognize that it might not always be coming from you. Oftentimes it is just because we are trained to think that way. In our Judeo-Christian framework, we have been trained to think in a dominator model with a patriarchal system over it. We don't think of ourselves so much in concert with the plants anymore, as I think we did back in the old, superstitious days of the Middle Ages. There was a lot of superstition, not science. There wasn't anything we could pin it on. It was all very mystical. Those same messages still come through the plants. It is just that we don't interpret them in the same way. One of the things that I encourage people to do is to recognize or to open themselves up to that communication. To me, it is pretty essential. That sense of opening up to something greater than what you are. It doesn't matter what you call it. Words get in our way. But spirit is something that a lot of people will accept. They won't accept god or goddess or Great Spirit or Allah or whatever. People will accept the word *spirit*. It is a matter of allowing people to recognize that they can flow into something that is out there. That they don't have to dominate the plants. They can be in harmony with the plants. The plants are our allies. There is a partnership going on.

COMMERCIALIZATION OF HERBS

I think that this herbal renaissance has become a primarily commercial interest. There are people making a lot of money now that started out in their garages. I remember when they were just making tinctures in their garages. So, commercialization has had that effect. Those folks have become more materialistic, clearly they have. When it comes up on *60 Minutes* and the cover of *Time* magazine, then you know that you are moving from the spiritual to the material.

I always believe, however, that there is a balance that occurs. I see that every year now at herb school. There was a decline in the early 1980s. There is a resurgence now. Without any advertising, we always have a tremendous waiting list for the school. I am always very surprised that somewhere down in the bowels of this renaissance, these young folks have been affected by what they have heard and what they have read. They come . . . to the school with a beautiful spiritual vision. They are not a whole lot different than we were at that age. I see them as being open to spirit—as open as I was at that age and as open as I am now.

WHY THE HERBAL RENAISSANCE IS HAPPENING NOW

There are several ways of looking at the current herbal renaissance. One might be that as there are cycles to everything, that we have cycled ourselves so far away from plants, and now we are cycling back. It is sort of a natural progression. When you get too far away from something, you have to go back. That is a clinical way to answer that or to look at that. We have been in a patriarchy for so long that now we are cycling toward a matriarchy, slowly moving back to that over many thousands of years.

For myself, I believe that the plants are calling to people. I don't think you have to believe that to find yourself called to herbalism. I guess that I would say that plants are *always* calling people. They always have and they always will. It may be that in certain ways this cultural climate is right for us now to hear that calling. It started with the whole back-to-the-land movement in the 1970s. I think the folks at Findhorn made it very clear that plants are always there waiting whenever we find it in ourselves to listen.[2]

It was just a matter of time until the medical establishment racked up enough of a reputation and a record of abuse. It is not necessarily abuse, but . . . that model has gone about as far as it can go for some people. I think also that enough time has passed since people had an original fascination and awe of the scientific method, and enough time has passed that people no longer have an awe experience of it. Just like television. You have an experience of awe of something, and after a while, it becomes commonplace. When it becomes more familiar, you can see more of the flaws. I think that is what has happened with the scientific method and high technology. That is a natural process. People then begin to turn back to something that they have ignored for a while. Again, some of it is cyclical, I think, but I would lay some of it at the feet of the medical establishment. If you read the history of medicine, there was clearly some abuse. There were real power plays and some real disregard for the body's own healing mechanism.... There was also a disregard for the plant kingdom, and all it has to offer to us. There was fascination with all things scientific and rational.

OPENING UP THE MAINSTREAM TO HOLISTIC HERBALISM

Many times I notice that if you are talking about people who are very deeply entrenched, it is only some kind of a crisis that makes them realize that herbalism is a valid healing art. Either it is a health crisis or a spiritual crisis or an experience—some kind of experience that shatters their belief system in some way and opens the door.

I don't think there is anything more that herbalists can do to encourage people to consider herbs. We are doing the right thing essentially. What I see around me is very positive. The creation of the American Herbalists Guild [AHG], for example, was a good development. It created a sense of excellence, striving for greater excellence among practitioners, and [it provided] a forum for herbalists to share and achieve more together. Those are all important things. The development of the American Herbal Products Association [AHPA], experiences where people band together, the idea of strength in numbers, all of these developments have been very positive. We are doing the right things.

Another thing would be United Plant Savers. What a wonderful development that was. Every aspect there has been a positive development. We need them all. Somehow they have arisen.

STEPS AN HERBALIST CAN TAKE TO PRESERVE HERBALISM

Don't lose your vision. It is difficult not to lose your vision. Like the folks I was talking about before, those folks that started an herbal business in their kitchen, and now they are millionaires. I am sure they have lost their vision. They have lost their initial belief. They have gained others. Some of them are positive; some of them are maybe not so positive. In whatever way, allow yourself to continue to be reminded of what is important. Sometimes you have to make a big mistake. It comes back to that idea of crisis. Sometimes, you have to make a big mistake in order to find the path again. I don't think mistakes are a bad idea. An herbalist has to keep taking risks, keep being willing to make mistakes. You have to be reminded of what it was that originally drew you in.

RITUAL

The very nature of ritual is that it can be anything. I do know, however, that some people feel that ritual has to be very formal and very serious and lengthy and involved. They feel that rituals must have a particular format and go a certain way. Certainly there are good reasons for that, but my view of ritual is a very broad one. A lot of what we do [in our everyday lives] is ritual. Certainly some of it has been diluted. Some of it has been weakened to the point that it really isn't ritual anymore, simply because it doesn't have the meaning behind it.

I would say that even walking among the plants is ritual. Take something that simple. It is something that can be either imbued with meaning, or it can be something that is simply mundane and very meaningless, spotless, without purpose, or without consciousness and mindfulness. Almost anything can be a ritual if it is imbued with that meaning. By the same token, ritual can be really hollow. I have been in plenty of rituals that were—in fact I have led a few, that just never really went any place. The spirit wasn't really there; the magic wasn't really there.

REDISCOVERY OF RITUAL IN AMERICA

I don't think this is necessarily directly associated with the herbal renaissance, but along with the herbal renaissance there has been a renaissance in a lot of other ways, too. There has been a great spiritual renaissance. So, ritual has been rediscovered in some ways. Some people for whom ritual became too formalized and lost its meaning can't ever really get back. It is lost for them in certain ways, but I have also experienced ritual in its capacity to awaken something within people that they couldn't have had awakened any other way. Certainly, the whole idea of plant spirit medicine falls under that category. I have seen ritual really turn people off, and I have seen ritual awaken people. I think it is a highly individual matter. It is hard for me to speak about ritual in that way. It is not something that you want to drag people into. Again, it is part of what you were saying earlier, you are led into things and things call you. You don't really know why. I have used herbs for ritual quite a bit. There have been times when the herbs have really brought their power to the ritual. It has been really palpable. I would have to say that there is really a connection there. I have been very privileged to have facilitated that a lot.

WHAT MAKES RITUAL MEANINGFUL?

Sometimes when I take folks into the garden, we will just stop for a moment. Instead of just walking into the garden, I will stop folks and say, let us all pause here for a moment and open our awareness. It will cause people to see and to sense things that they wouldn't have. It is a very simple act. By the same token, making a cup of tea can also be a very mundane activity. It can be hurried and rushed. It is almost like, people appreciate being led in ritual if only to slow them down or to give themselves permission to take a little extra time or extra consciousness around the action. People can make tea as a ritual. It is wonderful when that consciousness is there. It is my favorite way of taking herbs. It is warm as it goes into the body. The act of preparing it. The act of actually being in contact with the herb itself. The act of needing to take the moment to sit and drink. [That is ritual.] You can interrupt that at any point. And people do. I have seen people have a hurried cup of tea and drink it as they are driving their car or when they are walking around. Obviously, I do it myself. It can certainly be interrupted, but it is a ritual that I include in my class.

RITUALS THAT ANYONE CAN DO

The simple act of taking a moment to gather your consciousness before you head into the garden can be considered ritual. Another one for when you are in the garden is as you are approaching the plant, and you are ready to work with it, whether it is transplanting, we call it "conscious transplanting," or pruning or whatever, is just talking to, communicating with the plant. Some people have a hard time with talking with the plant. There are always students that say, "I can't do that." Most of them are really open to it, but there are a few who think it is really silly. I will make it really simple for them. I will say, "You don't have to talk to the plant if you don't want to, but listen. Calm your mind. If you approach the plant, approach it in the spirit of friendliness, in the spirit of cooperation rather than the spirit of 'I am doing this' as though it were simply an object. Think of it as another being." Most people can do that one. So, there is a greater sense of consciousness. That is ritual even though I don't present it as ritual. It is ritual because it is taking more care. Being conscious with an act.

PLANT ALLIES

"Plant allies" means something different for each person. I think mainly it is a matter of opening yourself up to the plant so that you are available to hear what it is that comes from the plant, if anything at all. You really can't expect so much . . . the plant will speak to you in different ways. Sometimes you just don't hear it. I guess that is the best way I can say it. It is more about your relationship with that plant. You can be in alliance with a plant if you have an openness to the plant cooperating with you in the process. Then you can open yourself up to a way of seeing that you haven't experienced before. You can allow yourself to have a cooperative relationship with the plant. You can allow the plant to work for you, or you can open yourself to the plant teaching you formally or informally. Let's say you are working with a client, and the plant acts in a way that you don't understand, instead of rejecting that, one way to be an ally of the plant is to keep yourself open to all of the possibilities. When you do that you allow the plant to become your ally.

The crucial thing here is that the plant cannot be your ally unless you allow it to be. We are the ones that seem to have the greatest influence on this planet, at least in the ways that we can perceive it. It is up to the individual to make that relationship. I mean, it will anyway, but you won't have much of an experience consciously.

EDUCATION FOR HERBALISTS

I have been taught in the classroom setting for the most part, although I did have mentor at one point. I had rich experiences in the classroom but it wasn't the same as informal teaching in the garden. It was a degree program. I don't know of very many other people who have the degree that I have, which is a master of herbology from a university. The MHs that you usually run into are master herbalists. So, I actually had a college-level master's program that was only initiated because this particular guy, a doctor of botanical medicine who I studied with, was alive. He is no longer alive. If it hadn't been for him, this program would not exist for the time that it did.

After the classroom setting, I would do individual work with him, but it was certainly not what so many of us yearn for, having somebody who transmits her wisdom to us through experiential teaching and learning. That set up isn't common anymore. On the positive side, it means that more people are having an herbal experience than ever before. It also means that in certain ways there is a lot more knowledge about herbalism. You know, you don't look at the village Wise Woman anymore. But then, on the other hand, I think that at certain times everybody had a little knowledge of herbs. You just couldn't help it. It just passed on by osmosis, almost.

I have a daughter who is ten now, but I can remember when she was in preschool. She knew what plants you could eat. The other kids had been taught by their parents not to eat any plants. I remember realizing that in our society all plants were viewed as poisonous by these suburban mothers. That kind of thing I don't think was present years and years and years ago. We were closer to the Earth and the knowledge of plant uses was disseminated at least to some degree.

I do mourn the loss of the Wise Woman. There is a tremendous amount of herbal knowledge that was lost, that we can't . . . that was lost completely because of the Inquisition and even by the action of the medical establishment. On the other hand, people are trying very, very hard to regain it. There is a lot of good work in the laboratories. There is a lot of good work confirming some of the old knowledge. There are even new things being discovered. St. John's wort was never used for depression. It was a much broader picture, so it lost a lot of the details. Elecampane is another example. It has a pretty limited use today. This is true of a lot of herbs, but this is off the top of my head. Same thing with echinacea. It is fairly well misunderstood as compared to the true complexity of the herb.

BENEFITS OF CLASSROOM LEARNING

In a classroom setting, you tend to get a lot more diversity, assuming that you have a staff, not just one person doing all the lecturing. I know there are certain programs taught in classroom settings where one person does all the instruction. But, for example at CSHS, we have a number of teachers over the course of the nine-month program. Students are exposed to maybe thirty teachers. Something like that—twenty or thirty. People come through and teach a one-day class. Or a week. Rosemary [Gladstar] will come through and teach, and Ed Smith will come though and teach. Sometimes they will be scheduled in and sometimes they will just call up and say they are coming through. It is a very rich experience. In that way, you get a lot of different viewpoints. That is really positive. You don't just have one person teaching you [his or her] viewpoint. I would have to say that is an upside to having a classroom setting. You end up having to make your mind about things. It is a compilation of a variety of teachers. It is really rich. In some ways, you might end up being a better healer. It is hard to say . . . we should do a couple of studies [laughs].

INTEGRATING DIFFERENT EXPERIENCES
OF HERBS IN TEACHING

One of the experiences that the students have at the herb school is that they each have their own garden plot. They have to care for that area over the nine-month period. It is primarily focused on during the first semester. After that, they go into more of a foundation of therapeutics. They get a lot more heady information during the second semester. That is a great experience the students have, the experience of having their own garden plot. They get a lot of one-on-one experience that way.

In my gardening classes, I include different kinds of information. You have to in a certain way. The students ask you what this plant is used for. It is a combination of direct experience and more intellectual material. I really love it. I think it is the best place to be with the students, actually out there with the plants and feeling and touching and sensing and smelling the plants. It is a great experience. You just can't beat having that direct experience.

I question schools that don't expose students to growing and caring for live herbs. There are schools where the teaching is all inside the classroom. I question an education that has that kind of focus. That is a little dicey. That is going an awful lot into the head and not into the heart.

SCIENCE AND FOLKLORE IN HERBALISM

In terms of science in herbalism, there is an imbalance. There is glamour associated with the laboratory and clinical studies. The danger is in only relying on those studies and not believing anything else. I think a lot of people have gone that way to the point where they don't believe in something unless it is scientifically validated. That to me is a big mistake. I know that it is rampant nowadays. But there is no stopping it. Again, for myself at least, I have to take a positive viewpoint and to use that information well. Make it part of my whole vision. Scientific studies do carry power. I notice that when I am in the garden, and we are doing an herb walk, I will always say a little bit about the uses of the plant so that the students can get a handle on it. I will mention some of the clinical tests that have been done. Studies carry a lot of weight. People will remember them. They believe it. Scientific method has become our god. Whenever that happens, whenever you do that, you run the risk of ignoring the magic. I think that is where education is key. I think that at CSHS we have a tremendous balance. As the need for validation of herbal medicine grew rampant, our director, James Green, caught it early on. He said, "This is not good. I am going to begin to focus the curriculum more and more on the gardens, on the herbs themselves as living entities."

HERBALISM IN GENERAL

I find that the world of herbalism is incredible. It is steeped in the plants which are by their nature protective. So, in certain ways you just can't go wrong. In the long run, we can't go wrong because sooner or later the plants will correct us. We are at least attempting to tune in to that greater force. So, at least when you try, it will respond. I consider myself very lucky to be here in this place and time. It feels right. The rewards in working with medicinal and culinary herbs are not monetary, but your rewards are the things you pick up over time being outside in nature, not having to be under fluorescent lights, not having to deal with politics and a boss. You get a reprieve in the world of herbalism. It is a great blessing. It is a great place to be.

Chapter 12

Kate Gilday

Kate Gilday is a clinical herbalist, flower essence practitioner, and woods woman. She tends gardens and teaches at her home in the Adirondack region of New York State as well as at herbal conferences throughout the Northeast.

Kate also cares for the wildlands that surround her home. Along with her partner, Don Babineau, she ethically wildcrafts medicinals for her own use as well as for some of the smaller herbal extract producers. Kate and Don have also created a set of flower essences from endangered medicinals and have begun using them in clinical practice in place of the tinctured material. The research is ongoing, but, as Kate relates in the interview, she is seeing success in using these flower essences to effect physical change in her clients.

Kate and I spoke over the phone on August 9, 1999, on the topics of endangered plants and ecological consciousness.

AT-RISK MEDICINALS

I hesitate to use the word *endangered* when I talk about medicinal plants, but I do think there are some plants that are at risk. The species I am most familiar with are the plants that are native to the Northeast. I have been an herbalist and practicing in the field, so to speak, being out with the plants, for over twenty-five years now. I have watched a few populations of plants in the Northeast dwindle. I do think they are at risk. Many of these plants grow in the North American woodlands.

I think what we have seen over the last number of years is a lot of the plants' environment being encroached upon by builders and loggers. I think at this point we also have to look at overharvesting of some of these plants. Herbs have become more and more popular over recent years, and people have been harvesting many plants from the wild—some of them with good hearts, but without an understanding of how the plant propagates or how to take care of the plants that you are harvesting. That piece is becoming less of a problem as people begin to be aware of the life cycles of the plant populations. Many trained herbalists and people who are studying with herbalists who are very in touch with that themselves, recognize that there are some plants that you need to be cautious with if you are gathering. Ten, fifteen years ago, we used to say, gather one-third to one-half of the plant population. We do not say that now about a lot of herbs. Now we say if you are going to take something, take a tenth of what is there. Make sure that you go back when it is time to pop the seeds back in the ground, and if you are gathering, make sure you are gathering so that you don't hurt the plant. Simply take a part of the rhizome.

COMMERCE OF AT-RISK MEDICINALS

There are people in the Southeast that I know of who are gatherers of the wild plants and have been for generations. These are people in Appalachia, people in the Smoky Mountains. These are people who harvest for their livelihood. It gets tricky at that point.

There are some "ginsengers" or people who are gathering black cohosh who know how to harvest so that they can return to the same place time and time and time again to be able to gather more. They know how to support the plants' proliferation. At the same time, there are people who know they can make a fast buck harvesting some of the at-risk plants, which get a high price on the market. They don't have that same attention or connection or commitment to the plant communities that the diggers I mentioned before have.

Many of the herbal product companies are taking steps to help alleviate some of the over harvesting. Herb Pharm, Gaia, and a number of other companies are growing the pressured plants that they will use in the future. They have begun plantations that will mature over the next number of years. Rather than harvesting from the wild, they have made a commitment to doing some real intensive gardening. I don't think that we know, and we won't know for a while, what the difference between wild and cultivated varieties will be in terms of constituents and components. Regardless, I think it is a really healthy way to go and an important way to go. Herbs are off the charts. People want to try these things because they have heard them once in a magazine or a news report or whatever.

EDUCATING TOWARD ALTERNATIVE PLANT USE

We can be educating people about other plants they can be using rather than plants that are pressured. The best way to solve this issue is through education. For almost every plant that is at risk, there are other plants that can support the same body systems or particular disease processes that can be used. They may be more gentle or subtle. You may have to take more of it for a longer period of time to see an effect, but I think in the long run, it is better to use those close substitutions.

ETHICAL WILDCRAFTING

I think that educating wildcrafters about the reality of the crop that they are harvesting is enormously important. We cannot simply say no one should wildharvest these plants. For instance, in some areas near to us, we have a local abundance of blue cohosh. Do we harvest blue cohosh in this area? Yes, we do. It is healthy for the plant populations that are there to harvest from the plants so that they can continue their growth cycle, their abundance, their reproduction. In that way, I think it is almost foolish not to. We tend to them like our own gardens. We are watching every year, watching the development of this bed. This one area that we harvest from is five acres of blue cohosh. That is a lot of blue cohosh. What we take isn't even a dent. We may take maybe six to twelve plants and we replant the crown. What we are finding in some areas is that the blue cohosh begins to die out because it is strangling itself. The same is true of goldenseal. If you have a mother plant, after a number of years you have all of these baby plants springing out from her, but her energy has been drained by all that reproduction, and then the plant itself dies from the center out. This is in contrast to the growth patterns of ginseng, which will grow for years and years as a single plant. These types of situations make the educational component so essential. By taking some of the smaller plants and some of these two- or three-year plants and replanting them elsewhere or using them in medicine, you are actually helping the health of the bed you are working in.

You have to have that commitment to tend that area though. As a wildcrafter, I feel committed to that. That is what I need to do to feel good about harvesting this plant, any plant in the wild. Even if it is an abundant plant, I still feel it is my obligation, it is my duty, my commitment, my love, to help them to continue their cycle in the best way that I can. Sometimes it is by gathering, and sometimes it is by propagating.

EVOLUTION OF THE PLANT-HUMAN RELATIONSHIP

Let's look at black cohosh for a moment, because if we hadn't started replanting black cohosh, my guess is that it would be gone in the next five to ten years. There would not be any more black cohosh. It is being dug out of the wild by the thousand pounds, maybe ten thousand pounds at a time. That is a lot of root. There is a great deal of woodlands in the East and the Southeast, but that is a massive amount of plant material. Now many larger herb companies are growing their own. They are doing all kinds of investigations and experiments in which they are growing the plant in the woods and they are also growing it in the fields. Researchers are watching to see how the growth works and how it changes.[1]

It is through these kinds of explorations that we will learn. We will learn from one another. It is exciting because it feels like what we are doing is we are trying to encourage and nurture these plant species in a way that has never happened before. It is a cocreative process. I think it really goes beyond the notion that *I* am going to farm, and *I* am going to grow this plant. It is more like working in partnership with the plant. We are asking the plant, what do you want and how can I help you.

We are growing into our birthright in understanding the connection between ourselves and the plant; it is incredibly sophisticated. It is incredibly deep and detailed. It is like a web. Everything touches everything else. Those of us who work with the wild plants in any way, whether we are growing them or harvesting them or using them in our practice, we are getting to know that complexity. It is the beauty of how the herbs work in the body. Unfortunately, it takes time and commitment to learn that. The reality is that there are a number of companies out there now that don't really care. They are really just making money on something happening in the marketplace that is lining their pockets. They are just there to make money.

ARE HERBS MOVING
DEEPER INTO THE WOODLANDS?

A couple of years back, we were asked by Herbalist & Alchemist if we could gather sundew, which is on the at-risk plant list.[2] What we decided to do the last two or three years was to watch the populations grow here and take a look at the wetlands in which they grow. They grow in this very interesting bog system that we have around here. We also decided to just travel more, to get out there in the field, in the woods, and to see if we can find more areas of populations. One question in a series that has come up for me is, Are certain things really endangered or are we not going out to where they are? Are they still around but in quieter communities that are deeper in the woods? Does that mean we should leave them alone? Or does that mean we can harvest, in a good way, a small amount?

We are still grappling with these questions, but I am happy to say that we have seen many populations of sundew that are very healthy and, in fact, just two nights ago we were out for stroll to look for a totally different plant, and the sundew was just right there just beginning to flower, this little tiny delicate flower, this little tiny plant. Prolifically. It was very exciting to see. We just said hello [laughs]. We didn't take any at that point. We are thinking about gathering a very small amount for people because in some cases there are indications for an herb where it is very specific.

I don't know if it is so much that the herbs are retreating farther into the woods as much as we only go so far off the trail as humans. Most people—not just the general public, but even the herbalists—do not go very deep into the woodlands or go beyond a certain geographical area in their findings. It is because most of us do use the herbs that are right in our own backyards, right in our own fields, very close by. But there are a few isolated herbs that really grow in a very specific environment that you have to find. I think that is really what we are looking at. They are the wild ones, and there is something about their wildness that is very untamed and at the same time very innocent. We take care with those plants. I hope none of them gets to be a star. Ginseng and goldenseal are proof of what happens when a plant becomes a "star." But now people are growing ginseng. That is very exciting. They are not making as much money as they thought they would, but it is still happening.

Molly Sheehan gave a talk back in Vermont about six years ago. [Editor's note: Molly Sheehan, a flower essence practitioner, runs Green Hope Farm.] Her connection with the plant world is way beyond. It is into the sphere of—you can call them angels or guides, spirits, devas—the real spirits of the plants. She spoke about some of the harder to find plants as choosing to leave, choosing to not just change form, but to actually leave this reality, this world as we know it, and feeling that their time here and what they were here for is done, is over. And not just the plants, but also animals and insects, fish, all beings, all creatures. My personal sense is that of loss because I tend to be someone who really does connect with the past. I find a great deal of sustenance from my roots in the past. I think that energetically the plants are still here but are here in fewer amounts.

We made flower essences from the at-risk plants.[3] We found their energies to be very powerful, much more powerful than the plants from which we had made flower essences that were not at risk. The at-risk plants are the real silent ones living deep in the woods. I am just watching the plants and being with them and using them as flower essences now. When I think of Molly's words, that some of these beings are leaving, it makes me profoundly sad. At the same time, there is a sense that they are leaving to make room for something else. Perhaps, in their leaving, it will become a major wake up call to human beings. Maybe humans will really pay attention. My hope is that that is where we are going.

DISCONNECTION FROM LAND

My partner, Don, and I have talked about this for hours at a time. We feel that people of North America, native people who have been here for thousands of years, this is their home. This is what they know. This is what their bones know. It is what their hearts and souls know. When they talk about ancestors, it is direct. It is right here on the ground; it is on the Earth. It is in these woods. When you sit with people of native descent they are totally at home in themselves, in the land, with each other, with all of nature; they just know this place. I feel like those of us from European descent have lost our connection to home. We have lost our connection to ritual that came from wherever we were. I mean, every culture had ritual, every culture had tradition, every culture had ways of celebrating the seasons and the Earth.

That got ripped away when we moved from the old land to this land, North America. The people that probably did the best were people that wound up on the same latitude as their former homes. At least they can connect with the climate and the seasons. How about people from southern Italy that wound up in upstate New York? Wow, that is different.

The land always keeps a very open trail. That is, it is very openhearted. It says come get to know where you are and who is here. We are seeing now the renaissance of people going back into native, indigenous cultures and trying to glean from them what it means to be connected. We are trying to figure it out. What is our land? Who are our ancestors? Where do we find "home"? I think we often find home in our own hearts, but at the same time I think that the land and the trees and the woods and the waterways can really be our home if we can just open ourselves to those experiences. There is such resistance. And fear. In our classes, we try to help remove that fear and separateness so that people will recognize the connection to the woods and to the native plants, the connection to this being their true home.

Part of it is shame and guilt. Part of it is the incredible shame and guilt from what our ancestors did to claim this land. They did it out of ignorance. They did not understand the incredible connection that the native people had with the land itself. They did not understand. European farmers came, people who understood how to tame the land, to treat the land as if they were the boss and the land was theirs to change. It was not a cocreative experience at all. What happened first was the forests were cut, the native peoples were destroyed, and their land was claimed. We need to think of the herds of buffalo. There were between sixty and eighty million buffalo when the West was first being discovered by the white man.[4] We now have seventy million dogs in this country [laughs]. We have replaced buffalo with dogs. I think there is a lot of shame that we carry, and guilt and pain, to think that is how we came here. It is very unconscious, but it is there. We need to realize that we do not have to hold that guilt; we do not have to hold that shame. We only have to open our hearts, and the forests, the plants, where we live will just rush right in and fill us up and carry us home to ourselves. It has happened for me; it has happened for Don; it has happened for my family; it has happened for people that we see. It is so freeing to finally release that cloak of disconnection that has happened from our own internal holding.

DOMINO EFFECT OF AT-RISK MEDICINALS

People have known that goldenseal is at risk for years now. I am going to say really for the past five to seven, or even ten years, people have known that goldenseal is not abundant. It is not something that grows easily in your garden. The thought was at the time, depending on where you live, you should use the berberine-rich plants where you live rather than using the goldenseal.[5] In the Northwest, it was Oregon grape root. In the Northeast, it was gold thread. Down South, it was yellow root. These were the native plants of those areas. People started using them instead, and lo and behold we used too much of them too. The thing about these plants that you are looking at is that most of them have a very slow growing period. Some take a long time to flower. Oregon grape is a good example of a slow-growing plant. Gold thread needs its own particular climate and environment. They are all native plants, and they all need something specific. Yet we are harvesting them, and they are taking "too long to recover." In talking about the berberine-containing plants, we should mention barberry. We have two kinds of barberry, American and Japanese. The Japanese barberry is also a source of berberine, and we are able to use it in many of the ways you would use goldenseal. My feeling is there are plants of European descent that are often easier to propagate, and they seem to do pretty well in North American soil. Those would be the ones I would like to see being used with the native plant being used only occasionally or very, very rarely.

A "NEW" WAY OF KNOWING MEDICINAL PLANTS

Matthew Wood is one of my main teachers right now. He incorporates more the essence of the plant in his work. Certainly he knows the constituents and components of each plant, but he concentrates more on the whole plant. He uses the doctrine of signatures among other techniques to help him learn how a plant can be used. He listens to the plant. I am sure that he has many wonderful conversations that happen on the internal level all the time. He uses this intuitive and magical way of knowing in dealing with plants.

We have to get away from this scientific view of "this is for this" and "that is for that." This component you need for this and this polysaccharide for that. Instead, we need to appreciate the wholeness of the plants. When you do that, you can perfectly match a person and a plant. There is no place that we can say, "Body over here, spirit over there." But when the human being, when the expression of that human being is blocked, when the life force or the energy is blocked and you can figure out the right plant for that person by knowing that plant intimately and getting to know that person intimately, you can make that little connection. Then all you need is tiny amount of that plant to really shift that person's healing in great ways.

The problem is that it is not something that you can market. That is where we need to have well-trained, well-developed, committed people who are practitioners who can sit and listen to stories, spend a lot of time out in the field with the plants, and can really put the two together.

HERBS IN THE MAINSTREAM

People are still looking for the quick fix, generally speaking. Until people are ready to make a commitment to their own health and to making changes in their diet, lifestyle, all of that, in their own healing, then the herbs are only going to go so far. Many questions also come up for me around the commercialization of herbal medicine. Who gathered the plant? Did they gather it with respect? Did they just take a big mower and mow it down and pop it into a bag? The tincture maker? Did they shake [the tincture] daily? Did they say prayers when they were making their medicine? Is it just in a lab where everyone is isolated from the plant material? This is where the idea of supporting your local herbalist comes in very strongly. That makes a difference.

Basically, it is up to the person. Initially, it would be up to their trusting their own inner feelings about what they are doing to heal themselves. They need to not be misled or not be thrown off by people who may have their welfare at heart but who tell them things that throw them off. In other words, "You are taking that herb instead of that medication! What are you, crazy?" That can definitely get in the way, that initial fear. People need to let go of the fear and allow the trust to develop. I have watched it with so many people where it does develop. What I mean by it developing is that they begin to trust the medicine more and more. Every time they go to the store, they will intuitively choose another product that is actually better, a better quality. I think there is a little silent whisper depending on the quality of the product. I think that can make a difference.

In terms of people really understanding that health is much more than you will find in any bottle, it is a struggle. I would love to see people use food for their medicine, and the air that they breathe, and the waterways, and time off to take walks in nature. Those are the ideal ways of healing. That is where we can continue our health or support ourselves when we are not feeling well. And by finding those things that give us joy. I don't see much joy in most people that I treat. One of my first questions to people when they come in for a consultation is, "What do you do to have fun? What do you do to have joy?" They just look at me like, "What?" Like that is such a foreign question. "What do I do for joy? Let me see. Well, my grandchildren give me a lot of joy." That is real. Then they go, "Oh, I really love reading Harlequin novels" [laughs]. "I have this one friend that I just love to sit with and we have a cup of tea. I haven't seen her for three weeks now. I should give her a call." I think ultimately that is where we want to guide people in the work that we do. I don't mean just me and Don, but all of us who are caring for people.

We want people to be just as happy and vibrant as they can. I think that the best way I have discovered so far is through the use of plants. They are happy beings, especially when they are flowering. Wow, what exultation is out there. No matter what plant species is flowering, you just go out, and there is an incredible amount of energy. When people are sick, what do we do? We send them flowers. It just raises the whole vibration. If we all walked through Deb [Soule]'s garden once a week, forget it, we'd never be sick [laughs].

MAKING AT-RISK FLOWER ESSENCES

When we first decided to make these particular flower essences, the at-risk essences, it was more that we were coming into communion with the at-risk plants and we felt that they were asking us to do this, the plants themselves or Nature, herself. It is kind of hard to explain because it is a little esoteric or "out there." We felt moved to make these essences. We felt that one of our roles as herbalists was to be a voice of the plants in whatever way feels good or right. Our first thought of making the flower essences was to just make them, to sit with the plants and glean from them what we could. What I do when I make a flower essence of any plant is to just go to the plant and sit and be with the plant. We call that journeying with the plant. I allow myself to get to a quiet space and open my heart. It is not like I hear a voice that says something grand. Little people do not come running out. I just have these subtle intuitions and these subtle thoughts that over the years I have begun to trust as not coming from my own left side of the brain, but they are coming from someplace else, and I trust that.

As I have used the essences over the years, I have found the information gained by journeying to be true, and it works with all sorts of people. The at-risk essences are still in a research phase. I wrote out descriptions for the plants. Interestingly, all of the plants that we made, we have years and years of experience just being around, not even using as medicine, but just simply being with them, growing them, being with them through all their stages, really knowing them well, and feeling a real kinship with them. At first, I just wanted to make the essences themselves, to make them available because my fear at the time was that the plants were leaving. Then what would we have left? That incredible energy, that incredible vibration, that wholeness of this land where I live would be gone. Have you ever been in an old growth forest site or a virgin forest site? It is different. It is powerful. There is a vibration or energy there or whatever you want to call it, that is very true and very old and very sustained and very healing. I found that to be true of the plants that we made essences from, too. Then we thought, Wouldn't it be cool if we could use these essences rather than the plants themselves?

Meanwhile, just backtracking a little, I didn't take the flower and put it in water the way most flower essences are made. I dropped water over the flower into the bowl of spring water underneath because I have tried all of these different ways of gathering the energy from the flowers, including these hysterical contraptions I used to make with coat hangers to try and bend the plants over into the water, but they all came out soggy. I began just using the drops of water this time. That feels really good. Trying not to disturb the possibility of a seed happening, we left the flowers alone.

We thought, Wouldn't it be great if we could use the flower essences instead of the plant extract when it would be indicated? We have been doing some experimenting with that. I am very happy to say that with a few of them we have found it to be very effective. American Ginseng's vibration on an energetic level seems to be very similar to its actual physical modalities. In other words, the flower essence helps with energy support, healing the endocrine system, helping [people] feel that they are energized if they have been in a place of depletion, and to fill them in a way that feels very nurtured, a little bit a boost of energy. We are watching that and seeing some people have wonderful experiences with Trillium essence for the passages of birth and rebirth from menopause, for women who are postpartum.

HOPE FOR THE AMERICAN HERBAL COMMUNITY

My dream is to find a way that the herb world can stay intact. That the people, whose heart and soul are really in working with the plants and appreciating all that they share with us, can stay in communion with one another. There are several places that we are beginning to see rifts. They are just small rips in the fabric, but my prayer is that we get beyond ego and beyond fear and having to have the last word and truly work in the name of the plants. That is my prayer.

Chapter 13

Rosemary Gladstar

Rosemary Gladstar is the founder of the California School of Herbal Studies and the United Plant Savers, and cofounder of Sage Mountain, a retreat center and botanical sanctuary. She is the author of the immensely popular *Herbal Healing for Women* (1993), the *Sage Healing Way* series, the *Gladstar Family Herbal* (2001), and coeditor of *Planting the Future* (2000). She has taught herbology extensively throughout the United States and led travel adventures to study indigenous healing systems in many parts of the globe. Her experience includes over twenty years in the herbal community as a healer, teacher, visionary, and organizer of large herbal events.

I spoke with Rosemary in the garden at Sage Mountain on July 30, 1999. I asked, in particular, about her experiences teaching herbalism and environmental awareness.

EXPERIENTIAL LEARNING OF HERBS

I grew up on a farm in northern California in a very beautiful rural area that was just full of wild plants and cultivated plants and gardens. Our neighbors and our community were farmers. In a very organic way, I was surrounded by plants as a small child. Not always, but often you will find that what you teach children to love when they are small, they grow up loving.

My grandmother had a big influence in my life. My grandparents were Armenian immigrants who had come here during the Turkish invasion of Armenia at the turn of the [twentieth] century. They had survived the genocide of the Armenian people. My grandmother and grandfather had been in the death march and had escaped. I grew up hearing these tales, but with one predominating theme: it was my grandmother's belief in God—she was a deeply, deeply religious woman—and her knowledge of the plants that saved their lives. That was a pretty strong influence on me growing up.

My grandmother used plants. She would go out and pick wild herbs. She used plants primarily as food, and as medicine only because that was the familiar medicine to them. A lot of the ethnic groups use plant medicine because that is what they genetically understand—it is in their blood! It is also affordable. My grandparents were very, very poor, as you can imagine, when they first arrived in this country. So, I have had this deep love of plants from the time I was very, very young. It has just been a lifelong love affair. My grandmother was my earliest influence.

I grew up in Sonoma County during the 1960s back-to-the-Earth movement, which was all about reclaiming our natural heritage with the Earth. I had already grown up close to the Earth, so for me going back was easy. I went back to the wilderness. I went up to the Pacific Northwest and the wilderness areas of Canada. I spent the next four or five years of my young adult life backpacking and meeting lots of elders, incredible native people who lived in the woods or were homesteading and farming and who, again, were very knowledgeable of plants as a lifestyle, not as an alternative. That continued to stimulate my love of plants.

When I was in my early twenties, I had a dream in which I wanted to ride back to Canada on horseback through the Pacific Crest Trail. I never fulfilled that dream, but I did end up buying horses and riding four and a half months into northern California. I went back to California to work to earn money for this dream. I opened up my herb store. That was in 1972. I really started my work with my community at that point. I didn't know a lot at that time, but there wasn't much known about herbalism in this country, so I knew a little bit more than my peers and a little bit more than the community. I opened up my herb store in the same community where I had been born and raised. I had gone to school or church with these people. There was a basic trust from the time I opened that store even though herbalism wasn't very popular. The herb store was really like a clinic and a lab and a college for me, all in one. I started the early conferences back in 1974. That is when I started networking with other contemporary herbalists, most who were in their twenties. There were a few elders. I started learning from them. So that's how it went: first my grandmother, then the plants, then my friends, in that order.

TEACHING ABOUT HERBS

I offered my first classes because very few people were teaching about herbs, certainly nobody in Sonoma County at the time. There weren't any other herbalists or herb stores or classes. In most parts of the country, herbalism was fairly devoid in the landscape. Not the plants themselves, but the tradition of herbalism had gone very deeply underground. Even though I didn't know very much, I had a desire to share it. I always tell people this—it doesn't matter how much you know; what's important is to share what you do know. That's teaching.

A lot of times people are paralyzed to teach. They feel that they have to be very good presenters, or they feel like they need to know a lot. I laugh because every single well-known herbalist who teaches started teaching when he or she didn't know much. They certainly didn't know what they know now. You never feel like you know enough. There is truth in that old saying that as you get older, you realize how little you really know. You have to work through that. What matters is that you share what you do know. I look at teaching as a sharing. It is sitting out in the garden or standing up in front of people and sharing what you know—not what you have read, not what you have studied, but what you know from your heart. You can amend that with information that is outside yourself, but the core teaching should be what you know. You are always safe if you stay within those parameters. When you stretch out, you say, "I am not really sure, but I read this in this journal or I heard from some other teacher." In that way you are learning and passing information on, but you are being responsible to your source for it. Anybody who is paralyzed by the fact that they don't know enough will never feel like they know enough. It is natural to feel like you know less as you learn more, especially about something as multifaceted as herbalism. I always encourage people to begin their teaching experience in their local [retirement community] or the local school. Go out and show them how to make salves, teas, lip balms, creams, or a bath salt, or maybe do a little herb walk. Maybe not teaching little kids so much about eating the plants, but more about the plants' interactions or the endangered nature of the plants. If teaching is a calling for people, if it is one of the ways that the plant asks you to serve it, then I wouldn't necessarily be discouraged by not knowing enough. Just teach what you know.

The other thing is never to be afraid to say, "I don't know that." There are teachers I know who do seem to know everything. They retain facts and figures and quotations so well in their headspace. But I can't teach that way. I just don't hold that kind of information. Perhaps my memory is faded.... I think it has always been faded though. I admire that kind of teaching, but there are lots of ways to teach. You can teach what you know and what you have experienced, or you can just say, "I don't know that, but I read in this book or here is a good resource or this herbalist really knows that kind of stuff." Then you are in safe territory.

On herb walks, herbalists have their own tricks. You are out on an herb walk, and people think you know so much because you are pointing out everything you do know, but what you are also doing is avoiding what you don't know. At least I do that. What is probably a better approach on an herb walk is to invite questions or invite others who know something about the particular plants to share their experiences. It makes a better learning experience for everyone.

GOALS AND METHODS FOR TEACHING HERB CLASSES

What do I want to communicate to my students? I think foremost in all of my classes my intention is to give the students an experience communicating with plants so that their relationship is directly to the plants and to the plant spirits. I try to give the students a direct connection with the living force that resides within each plant. I want to create trust within the students to believe, to go straight to the heart, and then to trust their experience with that. If I can accomplish that, I feel I have done a good job. The rest of the information is important and necessary, especially in the contemporary American scenario, but it is not essential. It is really going back to the old ways, trying to reestablish the old ways that people taught herbalism, where you spent years working directly with the plants. You learned to communicate with them. You learned to see them as living life forces that were genetically encoded with intelligence and their own agendas, and also had the great ability to provide medicine and food and fiber and air . . . really our life support for other living species on this planet.

Each teacher has different approaches. Eliot Cowan uses shamanic drumming and journeying to connect to the spirit. His teachings are direct. My methods are not quite so direct. I have found a different way that I think works comparatively [well] where you engage people's minds in interaction with the plants or you engage them in an activity with the plant (which is what Eliot is doing when he is drumming and doing the journeying). If you simply work with the plants and be with them and create time in your life to be with them, then that magic happens of its own accord.

I have found this to be a pretty successful technique. I wouldn't say it is one hundred percent successful, but it works for many students. I would say that a great many of my students do make that connection and stay life connected. That connection is there whether you are aware of it or not. You could not be alive here on this planet if you didn't have a relationship on some level with the plants. The deeper and stronger that it is, really, the stronger your life connection is. That is why you see people who are involved in plants, whether they are gardeners or farmers or healers or herbalists, whatever it is, that they are working with the plants, and shamanic traditions, they seem to be so richly imbued with life force. They are!

TOLERANCE AND ACCEPTANCE WITHIN HERBALISM

All approaches are needed. We become so judgmental. People are judgmental about herbalism. There was an article written recently about me, a very negative article because I bring spirit into my teachings. I was doing a workshop at a hospital, and the author of the article felt that it was a very inappropriate situation for talking about spirit. I honor that person for her opinion, but I certainly felt that I had the right to speak the way I did, and all of the people who enjoyed it had the right to hear it. I would like to suspect that we have enough tolerance to include anything. Who is to say where spirit resides? To say that mainstream clinical practitioners don't include spirit is to say that we have the only handle on what spirit is. I think each one of us finds a way to create our own mythologies in this world that support who we are. To say that so and so doesn't have a grasp of spirit is to say there is only one image of God. I certainly don't want that shoved down my throat, or even suggested. There are many pictures of God, and many mountaintops where God resides, and many paths up that mountain. There are also many gods and goddesses.

It is the same with herbalism. Science isn't my way because I don't understand science, or I don't understand what I consider to be my perception of science. But I certainly don't want to pass judgment on any way that people want to use science as a teaching tool. It's necessary. It's here. It is part of the whole.

I make judgmental statements too. We all do. I would like to say that everybody should go out in the field and connect directly with plants; but quite frankly, I know a few really incredible herbalists who are as connected as anybody else, yet they very clearly state that they don't "know" the plants, what they look like, how they grow. They don't see the body of the plants. They don't "know" the plants, so they don't spend time with them; they don't need to walk out into the garden. And, yet, they are really incredible herbalists. It is not my way, but we need to make space for all approaches.

There are some herbalists who work in clinical situations devoid of plant material who have no understanding from my perception of how the living plants exist; I know a lot of herbalists who wouldn't want to sit out here in the grass because they might get chiggers or something, and yet they have an astounding connection with the healing properties of plants. So, I think that it is our judgment. Just like, maybe, there are some herbalists who would say that we touchy-feely herbalists talking about spirit and hugging trees, we are just a bunch of old "tree huggers." That to me is a judgment.

There is room for it all. We need it all. We don't need to be any more judgmental in this field than in any other field. But we are human so we will be. We think, *(sarcastically) My way is better, thank you. I know how to connect with the plants. My medicine is better because I sit and talk to the plants.* I am not saying this is true. I am just saying this is how we run things through our heads, just like the clinical herbalist might think, *We have tested these plants. We are assured of their quality. We have done thin-line chromatography, and we know that our medicine and what we are offering to these people is better.* It is just a judgment. What we need to realize is that it is all needed, and it is all necessary. I think that is the holy truth.

TEACHING THROUGH EXPERIENCE VERSUS LECTURE

When you are teaching you have to convey the plants to the people. You have to pull in their magic. I primarily do that through hands-on stuff. I am not really a lecturer at all. I never have been. I am not really the best speaker for lectures or keynote presentations. I am not compelling in that way. What I am very good at doing is working with people and conveying this message. I know it works. I can convey the plants through me to them. But when you are out walking in the woods or in a garden, you don't even have to work at it. You can just take people out, and they get it directly from the plants. So, that is easier.

CLASSROOM LEARNING AND APPRENTICESHIP EXPERIENCES

Well, there are advantages and disadvantages to both systems of learning. Most people would say that one-on-one is the best, but I am not sure that is always true. Some of it depends on the student. For some people, working even in a small group is a challenge because they tend to be shy or other people's energy overpowers them. They spend a lot of time thinking about other people's energy without being able to stay focused. I personally look at that situation as a really good teaching [opportunity]. When people tell me they don't work well in a big group, I think, *Well, you probably should come and try it out, work on it.* You don't have to work through it, but you can work on it. You find tremendous teachings in group dynamics that I love. You have the teachings of how the shy people learn to assert themselves or don't so they are always feeling left out. You have the people who are the attention-getters who have always been allowed to that. You see this tremendous dynamic. You have a great opportunity to create world peace every time you are in a large group. I usually suggest that in our seven months together during the apprenticeship class. We are thirty people or sometimes thirty-five, and we have the opportunity to be a typical group of people. There will be people that don't rub us right; people that we really connect with; people that we think are really off-the-wall. Let's see how we all fit on our boat together. How do we do on this little survival boat? My challenge always is to make each group, no matter if it is two people or thirty, work; that is, be successful for each person aboard. If I can do that, I have accomplished my task in creating peace on earth.

One thing about teaching in front of a large group is that you tend to be a lot more prepared. If you are one-on-one, you are often more casual. This is a little detail, but you will find that is true of most people. I watch very accomplished teachers when they come to Sage Mountain. If there is a large class—anywhere from ten or upward—there is a lot of preparation. There are notes. There are handouts. You get very organized so that you can get from A to Z knowing you are going to have time for questions later. A lot of students learn really well when there is that level of organization. I am not saying you can't do that with a small group, but I would say that if any teacher were honestly asked to evaluate how prepared he or she is to teach one person versus teaching twenty people, you are going to probably find that he or she will do more prep work with the larger group.

In a one-on-one situation, you have a more casual environment. You usually are less prone to be the teacher. If your student lives with you, [he or she sees] you every day. [He or she sees] you get upset with your family. [He or she sees] you not always eating perfectly. Or, [he or she sees] you eating perfectly. Whatever it is. Your relationship as teacher is not as strongly defined. A lot of people will think that they learn better in one-on-one situations, but my experience has been that oftentimes when you have a clearly delineated student-teacher relationship, people learn better. We know those relationship lines are very fuzzy; we know there is no clear definition of student and teacher, but when you have that structure, students tend to learn better. People become clearer about what they are coming to learn from you rather than becoming involved in your family life or asking you to become involved in theirs. There is simply more clarity, I think. Of course, there are certainly great teachings in this as well, just not always about herbs.

I find that if you are organized, you can give your students a lot of opportunity to learn. You don't need to get involved in their personal lives. I try not to get involved in my students' personal lives. If I see that they are in pain or are ill, of course I will always help. That is what some of the program is about—helping to create and maintain balance. If I see that they are having problems, difficult problems, with other students in the class, I might create some parameters to help that, but I don't necessarily enter into their working that out for them. I feel that I want to support them in their process but not necessarily solve their problem for them. Part of it is keeping focused on your educational objectives and how to accomplish them. I know it sounds very sterile, but my classes are not sterile by any means. What I have learned over the years is that there are some necessary guidelines that can help you accomplish your objectives in a class.

DISAPPEARANCE OF THE ONE-ON-ONE APPRENTICESHIP

If you want to learn blacksmith skills, there are a few people who will take one-on-one apprentices. Or if you live in a community, you might be able to hire up with a blacksmith who will teach you his [or her] skills; but usually you have to go to school to learn that. It is sad. I think it is an incredible model. I admire people that do traditional apprenticeships, but I have heard from the teachers themselves that much of the time the expectations are greater than they care to meet. You have somebody in your space a lot. We have so much in our space already. We have so much busyness in this world, and families, and responsibilities. Oftentimes you need more alone time. If you have an apprenticeship, a one-on-one apprentice or two people, they may feel that that this is their time with you. You may be feeling more like all you want to do is sit down with your feet up on the porch and look up at the stars. . . . So there are a lot of issues there, but I do agree with you. I don't want to see that tradition die out. I don't want to see that model of apprenticeship become obsolete. And hopefully it won't. I don't think it will. I don't think it is dying out. I just think it is beset with its own challenges. People have to be willing and flexible.

WHAT THE NEW HERBAL CONSUMER NEEDS TO KNOW ABOUT HERBS

I think that the very first thing they should know about herbs is that herbs, like everything else on this planet, are not an endless resource. If you are even going to tinker in herbal medicine, even start to investigate it, the very number-one first issue is understanding the plants themselves in relationship to their communities and native environments. If you are going to use plants, you need to become responsible to their health and the health of their environment. To me that would be the number-one thing. People don't want to take that approach because they are going, "Ahhh, we have to learn about dosages, and what's safe, and contraindications. Now I have to be environmentally responsible, also?" We need to have an educated public in terms of the environment and the environmental issues surrounding herbalism right now. That is my personal agenda.

Following that, I would recommend people not come to herbs for medicine as their primary interest but come to them from a more ethnobotanical perspective. Understand how plants create the air we breathe, and how they have been the basis of all fiber throughout history, and the bottom support of the food chain no matter what you eat, and also provide incredible beauty for us and relaxation. So, come to it as a way of life before you go, "Okay, now let's look at it as a medicine."

I would ask the American public to not view plants as a drug in a bottle in the local drugstore but as a way of life that has been supporting life for as long as it has been on this planet. I am not talking about going into a long ethnobotanical course in college. I am just talking about taking hikes out in the woods or walking in parks and looking at the plants, noticing how beautiful they are. Maybe reading a book or two on the interrelationship of plants and humans, so that you have this overall vision of what plants really are about, and that a very small part of what they are about is medicine.

Then, the next thing I would suggest is that the American public have an historical perspective of plant use, that they begin to understand that herbs are safe and very clearly understood. What we don't have a long history with in herbalism is how herbs interact with certain drug medications because drugs have only been here a very short time. A lot of the contraindications are because drugs are the newcomers on the block, not herbs. Plants and their interaction in human beings have been documented for literally sixty thousand years. Individual plants have been found associated with humans as far back as twelve thousand years ago. You can look at very old artifacts—ancient manuscripts, pictographs, and mummies that have been found—and see that humans have used plants for a long, long time.

So, we do have to educate ourselves about the interaction of some of our stronger herbs with drug medications. The herbs have been here for a long, long time. The newcomers here are these drugs. Maybe we should start to get some long-term perspective on some of the prescription drugs that are killing people off right and left and all of the heart medications that we are giving to older people even though we don't know their interactions with the other drugs that their other doctor is giving them for bladder or kidney issues. To me that is a much bigger concern.

INTEGRATING ENVIRONMENTAL AND ETHNOBOTANICAL CONCEPTS INTO HERBAL EDUCATION

When I go to suburban New York at the Omega Institute, one of my topics is ecological concerns of herbalism.[1] I may weave that topic into the medicinal uses of the plants because certainly that is why half the people are there. I think the environmental factor should be a part of every one of our topics, though. What I always say to people, every place I go, is if you are going to use herbal medicine, your first and foremost concern should be for the plant communities. We can carry forth the tradition of herbalism, but to destroy the plants that the tradition is based on is tremendously irresponsible. I think ecological impact should be interwoven into every topic, but the only way you will get people to listen to that message is to address what they feel is important—how the plants will benefit them. You weave it in there as part of the conversation and part of the teachings. I think that is important for any of us to listen to. It should be part of every conversation. So, it would be like, yes, you need to learn appropriate uses; you need to learn different preparations. Do you buy tinctures or do you buy teas or do you buy cough syrup or pills, and how much do you take? Do you buy standardized extracts or do you buy whole plants? Oh my God! It is so confusing.

All of that is important to learn, but the most important thing is you need to know about the plants that you are using—who they are and their place in the great web of life. Herbal medicine is not isolated from its source. Our drugs may be isolated. We don't even need nature to create pharmaceuticals, but you do need nature in herbal medicine.

After the interview, Rosemary and I talked a little bit about how she is always suspect of the easy answer to a question. She thinks there are some standardized answers to questions. One example is whether she feels something is lacking in a strictly clinical education. She feels it is her personal challenge to search deeper for an answer and to be confrontational, if that is required.

Chapter 14

James Green

James Green is a passionate supporter of folkloric herbalism in America. He is the former director of the California School of Herbal Studies where he taught the "Technology of Independence," skills to empower and detach oneself from market dependency. James also teaches ethical harvesting techniques and ecological herbalism. He has served on the board of directors for United Plant Savers and is presently part of its advisory committee.

James is the author of two books: *The Herbal Medicine Maker's Handbook* (2000), a technical guide to making folkloric medicines, and *The Male Herbal* (1991), one of the only comprehensive volumes on male health and herbal medicine.

I spoke with James over the phone on January 4, 1999, a blustery Vermont winter day (not so on his end in California).

HISTORY OF AMERICAN HERBALISM

About twenty-five years ago, I met a woman named Norma Myers. I was living up in Canada at the time and was teaching in the local high school. I had already been turned on to plants, but I wanted to know more. Norma lived with a Native American community up there. She was a medicine woman. I decided to follow her. She took me to the seashores and showed me the plants that grow there. She introduced me to the plants that grew inland and taught me how to cultivate them. She was a real folkloric herbalist.

I call Norma and the herbalists of her generation the "*winter* herbalists." They took care of the seeds of knowledge when herbalism went into the dark ages in this country. Herbalists like myself, Rosemary Gladstar, and Michael Tierra met the winter herbalists and were inspired by them. We are now caring for the seedlings of the herbal movement. In that way, I feel like I am a springtime herbalist. Our students, the ones that are learning from us, are the summertime herbalists. The herbal movement is blossoming now.

AMERICAN HERBAL RENAISSANCE

I think Americans are discouraged by the allopathic alternatives that dominate health care today. American herbalism that we see today started with the back-to-the-earth movement when people tried to become more independent. They started to get off the [power] grid; they started growing their own food and raising their own animals. People started to realize the value of small community groups. In general, there was a pull to get back to the earth. The back-to-the-land movement was an incredible catalyst for the resurgence of herbalism in this country.

The young herbalists of the time started saying, "You know, these plants that grow all around us can be used as medicines. Here's how you can grow them. Here's how you wildcraft them. Here is how you can make medicines from them." The medicinal properties of plants [were] rediscovered. Counterculturalists were really ready for that. They wanted to get away from mainstream dominance. They wanted to get back into their own sense of self and their own sense of independence.

Now that herbalism has gone mainstream . . . it's going back to where it was before we started. It's going back to the drug companies and pharmaceutical companies taking over herbal companies and marketing the herbs like drugs. It's gone full circle again.

People like myself are out talking to the public, saying, "Wait a minute—don't forget your independence!" I'm writing books now on how to make plant medicine, how to harvest herbs, how to grow them. I want to make sure people realize that herbalism is going full circle. If the herbal community is not careful, we'll end up being dependent on a marketplace again. That's my line right now; that's what the California School of Herbal Studies is about. I teach people to continue their independence.

USE OF SCIENCE AND FOLKLORE IN AMERICAN HERBALISM

As public demand for herbal products has grown, the marketplace has begun to harness the field of herbalism because there is money to be made. Many herbal product companies formulate and advertise using this pseudoscientific reductionist mind-set where they talk about standardization and active ingredients as the most important aspect of the medicine. You are now hearing the same foolishness about "doctor approved in clinical trials" that you hear with pharmaceutical drugs. Consumers are looking for alternatives to the pharmaceutical drugs, and they are ending up with herbs marketed to them in the same fashion. You can derive high-quality herbal products from that process, but with it comes the sense that an herb can act as a "magic bullet" of sorts, a one-step answer to the health issue at hand. When this happens, consumers start losing the essence of herbalism: to take care of yourself and prevent rather than treat.

At the California School of Herbal Studies, I teach students to make their own medicine in a folkloric tradition. Once you can identify the plants and understand the ecological impact of harvesting herbs, you can make your own medicine. It is easy to do. It is like cooking. I'm writing an extended version of *The Herbal Medicine Maker's Handbook* [Green, 2000] for the public. In it, I show how to use wines and vinegars as a menstruum rather than always using alcohol. Wines and vinegars are excellent menstruums and were used for many, many years. I try to concentrate on materials that are readily available.

ENDANGERED PLANTS

Right now, we're using a few herbs and overusing them. If this practice continues, we will lose those herbs for good. Herbs such as goldenseal and black cohosh are going into a very dangerous place, ecologically. Herbalists have now realized that. We always used to say, "Wouldn't it be nice if people started using herbs?" Now that more people are interested in herbal healing, we are really concerned about the integrity of native medicinal stands. The demand on some of the plant populations is too much. Now we are going back and saying, "Wait a minute, folks. Don't wildcraft indiscriminately. Concentrate on buying organically grown herbs, until we are sure that the plant populations are strong. Otherwise, we will wipe out the natural stands."

There's no evil here. No one is deliberately doing something harmful. It is simply an issue of enthusiasm and the exponential expansion that American herbalism is going through right now. It is the job of the herbalist to help people to learn about the situation with at-risk plants and offer solutions to the public.

I approach this issue in my teaching in two ways. I encourage people to learn how to cultivate herbs that they will want to use in their practice, and I educate students about local wild plants, which grow in abundance. You're never going to overharvest plantain and dandelion and other common plants like that.

FUNCTION OF "MEDICINE" IN HEALING

There are so many different interpretations of the word *medicine*. Herbal medicine works in large part because individuals go out and get involved with plants. They reunite their spirit with the spirit of the plants. By going out and gardening or wildcrafting, you are getting in touch with the plant populations. This can be a very grounding experience. I think that's a major portion of the medicine's properties. Once you get involved with the plant in that way, you may not even have to harvest the plant material in order to realize its benefit.

When you buy something off the store shelf, you're missing the personal interaction with the live plant, with the spirit of the plant. Herbalism is a relationship between plants and humans. That relationship becomes muddled when you simply purchase a tincture or bottle of capsules. When you begin your journey with a processed plant in the form of a tincture or whatever, you certainly get pharmacological activity. There is certainly some benefit from it, but to me you've only got about a tenth of the medicine now. The simple act of picking a plant, bringing it home, and making a tea out of it makes a far superior medicine.

In talking about medicine, you must also talk about the individual's belief system, that is, if a person believes an herb is going to work for them, then it's going to work for them. You basically heal yourself, and if something helps you to realize that you have that power, then that medicine "works" for you. If you believe in pharmaceuticals, then they are going to work for you. I don't look at one type of medicine, or one system of healing, as being better than the other; one is simply more appropriate, depending on your belief system.

The beautiful thing about herbal medicine is that it has so much aesthetics to it. It has so much life force to it. At its center, herbalism is so much about getting in touch with beauty and using all of your senses—touching and smelling and tasting and feeling. If you go to an herb store, pull a bottle of herbs off the shelf, and open it—and look at it and smell it and taste it—you've gotten so much more involved with that medicine than with a tincture which you just put in drops of water. The more you involve your senses in the process, the more you are enlivening yourself.

CORPORATIZATION OF AMERICAN HERBALISM

As herbalism in America becomes more popular and thus more lucrative, it becomes more prone to corporatization. Instead of talking about the reciprocity between plants and people, you begin to hear lines like, "Here's the magic bullet. Take ours, it's been clinically approved; take ours, it's the best; take ours, it's the most powerful." More powerful based on what? Based on the availability of a particular so-called active ingredient? Or, based on the heart space of the person who harvested the herb? Or, the heart space of the person who made the medicine? Those are very important vibrations as well. It is all so muddled in advertise-speak that it's just lost the flavor of herbalism. The primary flavor right now is that of a product being marketed.

Many of my friends who did cottage industries back in the day made the decision to expand with the demand. Those companies have become very big now. Herbalists like myself, who decided to stay small, are still doing cottage industries. There are a lot of us around; you just don't hear about us as much anymore because there's so much "shelf-space fighting" in the health food store. Large herbal product companies will come into the local co-op and offer the buyer volume discounts and great sale prices if the store will sell their product. The larger companies simply have greater financial means to offer deals like that to the stores. In this way, they just push the little guys, the cottage industries, off the shelf. The little guys are only selling in certain geographical areas now, where the public more poignantly understands the need for heart-centered herbal businesses.

The managers in the health food stores have a great responsibility when they decide what to stock on their shelves. I tell people a good herb store would have four or five different brands of herbal products on its shelves. Then, customers can come in and use their intuition to decide what product will best match with them. They will feel it. You give them a large array that should include small companies as well as large companies. Sadly, though, the stores often make more profit by just doing business with large companies that give them excellent sales deals.

MONOPOLY OF SCIENCE

There could be a person who loves to garden and still loves to read about sesquiterpenes and flavonoids. It trips them out. They like that point of view. If you can have that kind of person who enjoys organic chemistry and also likes to lay in the red clover and dream about plants and devas, then that's great. The problem is when someone comes along and says, "Oh, unless you understand the pseudoscience of it, you can't be a good herbalist. Unless the product has been standardized to a certain active ingredient, it can't be any good." That's where I have my problems. We get this quiet dominance that says, "Well, if you don't understand the language of science, then you're not good" or "it's not a good product." Students who come to my school think they have to understand organic chemistry, and they don't. If they choose to learn it as a component of their herbal studies, that's great. It is not required, however. You don't have to do a weight-to-volume tincture to do a good tincture. You can work with the folk methods and make a very good tincture. Healing with plants has been around much longer than the weight-to-volume tincture.

FOLKLORE IN AMERICAN HERBALISM

You will never be able to outlaw [folkloric approaches to herbs]. There will always be certain individuals who will work with herbs in a folkloric way, whether they are in the spotlight or underground. The only way the FDA has any power over herbal healing is in the marketplace. Within the setting of the marketplace, the job of the FDA is to make sure you have unadulterated products that provide the information that the consumer needs. On a large scale, you really cannot market a folkloric herbal product. For instance, in order to exist in the marketplace, you need to make a tincture based on a weight-to-volume ratio. This allows you to be certain of the strength of the tincture and therefore determine a proper dose for the consumer.

The folkloric method works best in the home or in the community, if you have a local herbalist. That person knows the plants intimately. They can use folkloric medicines because they have an intuitive sense of the plant's strength which allows them to dose people appropriately. Folkloric medicine making remains very viable when people work on a small and intimate scale. Those are the people I teach to.

NATURE OF HOLISTIC HEALING

[Holistic] can mean so many things. It depends on the person's point of view. I would be the last one to say that herbs are the only thing you need for health. An herb is just a part of a program. For people to be well, they have to be happy. They have to be doing what it is they want to be doing in their life; they have to be getting their rest, relaxation, exercise, and decent food. You can't use herbs to replace decent food. You take care of your spiritual self, your physical self, your mental self, your sexual self. So, holistic just means to be well rounded and balanced, and every person has [his or her] own balance that [he or she] must find. Holistic means getting to know yourself well, getting to know what makes you happy, what pleases you, what foods feel good to you, how much rest is right for you, how much exercise is right for you. You watch enough TV to entertain you but not too much, use computers if you want to, but you don't have to . . . whatever works for each individual. The beauty of it is that each person discovers [his or her] own unique individuality, rather than trying to be like someone else. And the diversity of individuals is what's so beautiful, so that's why I always say you can't say that one diet is right for everybody; you can't say that one herb is right for everybody; you can't say anything is right for everybody, because it never is. To me, that's holism.

I don't know what an average American thinks holistic means. Perhaps, they think holistic means "body, mind, spirit." That's good, but they need to understand that "body, mind, spirit" are not divided into three categories; holistic means that there's no division. The body is as sacred as soul, as sacred as emotion. We can treat everything as being sacred. Take anything as sacred; everything as sacred. When people get to the point where they can bring all of that together and start to really appreciate everything, even plastic. . . . That's holistic!

I think if a person is really holistic [he or she will] allow things to be. [He or she] will not gang up against a certain idea, or against a certain group because [he or she doesn't] agree. I really feel that the most enriching space to be in is that of compromise. If there are two opposing ideas out there, or two individuals with different ideas, I think that the creative compromise that comes from the meeting of the ideas is often the best. I feel that the pharmaceutical community and the herbal community are going to come up with something far beyond both of them in time. If we can get the sense of opposition, us against them, out of there and just sit and honor each other as individuals, we can see how we can work together as opposed to only for ourselves.

To say something is wrong tends to resist the flow of ideas, the flow of creativity, the evolution of time. We need to embrace people and ideas. We need to increase our knowledge and our feeling that everything is okay. Everything is wellness. There's joy and happiness inherent in the universe, and we are not losing it. It's there for us. We can experiment with things, and by experimenting it doesn't have to mean we're going to destroy it all. We have this doomsday quality about us that, if we play with this, we're going to destroy everything. I don't necessarily feel that.

POWER OF THE INDIVIDUAL IN THE HEALING PROCESS

My basic precept is that health and joy is our natural state of being. If we let go and let ourselves be who we are, we will naturally bob up to health and to joy. That's our natural state. People have ceased to believe that; in order for them to have that, they have to have something to help them. They have to have a medicine; they have to have a practitioner. They feel they need that. If that gives them the excuse to allow themselves to have their natural health which was always there anyway, then fine. Let them have it. Whatever it takes.

A lot of people do feel that they need a "medicine" to make them better. At a certain point of your relationship at the client level, that's an important thing to work with. You start by giving them the "medicine." You say to [them] if they feel they need the "medicine" to make them well, then they can take it. So they take it, and they find themselves getting well. Then, the practitioner's job is to help [people] realize that they really didn't need that herb, that the healing came from within themselves. The healing was their own spirit rebounding to allow themselves to have the life that they want. They didn't need the medicine. They didn't need the practitioner. Then, you start to empower [them] to realize their own inherent ability to heal themselves and keep themselves well. You've helped them realize they don't need to rely on someone or something else.

Many people have to go through steps to understand that they have the power to heal themselves. In the beginning, it means little to them to say, "Everything is fine and well and the natural state is health, so go out and be healthy," but that won't be enough for them at that point. The herbalist needs to act as a coach and guide them to this awareness. When they are in a stronger place and they have experienced their revised health, then you start to teach them that it was all about them anyway. They just needed an ally for a while.

THERAPEUTICS IN THE MARKETPLACE

There are obvious risks in seeking therapeutic help within the marketplace setting. That's what the FDA and the AMA [American Medical Association] try to control. As an herbalist, you're not supposed to diagnose and prescribe. If a person comes in [to the local health food store] and has [a] diagnosis from a qualified practitioner already, you're still not supposed to prescribe.

I also feel that it is important for a health food store employee or a community herbalist to recognize the limits of his or her knowledge when answering questions or giving advice to someone in need. As an herbal practitioner, I refuse to be licensed. I refuse it because it limits me too much, but I would never treat beyond my knowledge. I would always say to the person, "Well, I don't know," or I would suggest see[ing] someone else who might have a more specialized knowledge than I have. People, especially those in the marketplace setting, just need to realize that they must be very straightforward in what they do and do not know.

If I had a store, I'd probably be asking the people who work with me to be talking much more about health than disease. I try not to talk about disease; I talk about health. If a customer came in with a question, I would just try to give [him or her] herbs that I know are very supportive. I think that if you are going to have people on the floor giving health advice, they need to have some kind of a model to work with, Ayurvedic or Chinese, or some kind of a model, some kind of a constitutional model.

I know the American Herbalists Guild talked about training that very group of people [i.e., natural food store employees]. It is a part of the herbal community that really touches more people than anybody else. At this point, I think when therapeutics come up in the health food store, customers need to be referred to books. I think what you should have is a good library—books written by practicing herbalists, not these people who have been plagiarizing other people. There are so many books out there now.

When I talk to [people] about [their] health, I don't talk about an herb until a lot later. I talk to them about themselves, about their life, about their happiness. What do you want in your life? If they are not getting what they want, what are they going to do to get what they want? I don't start talking about herbs until way, way later. Then I'll just give them something to take with them—tonic herbs that are not treating the symptom but are just supporting the person. At the store they're trying to sell products, and there you get your problem. If they try to sell product, they are going to try to push them to this or that. And a lot of stores do that. You have conflict.

If I worked in the marketplace setting, I would spend a great deal of time directing people toward tonic herbs that would meet with their constitution in a harmonious way. I would train my employees to speak with customers about wellness, and then I would have a model they could use to sense the person's constitutional nature and turn the customer on to tonic or supportive herbs, something with strong nutrients that the person's body could thrive on. I wouldn't be treating symptoms; I would be treating [people]. If you get to the right tonics, and [people] get their [lives] working for them, those symptoms disappear. The symptoms are reflective of a resistance in their own [lives] to the natural flow. That's holistic, see. I would never talk to [people] based on a symptom. They would come in with that in mind, but I would divert the conversation by saying to them, "How's it going? What do you want?" This is the most primary question, "What do you want in your life?" because it is the fact that they are not getting that or have not defined that which brings up the symptom. So holistic medicine doesn't sell a lot of product sometimes. There's the problem we have. We have people calling it holism but they're trying to sell products. I could sell a person a lot of herbs, but I wouldn't do that until I sat down with them and talked to them about themselves for a good half hour.

My teacher, Norma, would help people to regain their sense of health by getting them out in the garden. She would get them outside. They would be digging beets and juicing beets. She would have them go out and find their own medicines. What she was doing was diverting them from thinking about the disease, and getting them into a life activity—out into the air or out into the garden, the light, the sun. She'd get them involved in life and they would quit talking about a disease. It was so powerful. That's what Norma did, and that is what I would do as well if I played a role in the marketplace. I would get the people talking about their [lives]. Get them back into some uplifting ideas about things they want to do, and then encourage them to do that. I would give them some tonics and a little bit of diet to carry them along with that.

I would love to be the person who trains the employees who work on the floor. I think it's really important. I don't think there's any of that being done. Most of the training is either done by herbal product companies that want to sell their particular product or by the store management that lives by the sales in that department. It gets interesting when the drive to sell a product interacts with a person's need for health and healing. If someone comes in talking about his or her health, there's so much more to it than a product to buy.

ENDANGERED PLANTS, MONOPOPULARIZING, AND BIOREGIONALISM

Monopopularizing is kind of like the cousin of magic-bullet consciousness, right? They are doing the same thing—here, take this herb and it'll do this. That's all: that reductionist, magic bullet, treat the symptom with this thing. That is what is being marketed; that is where the marketplace is. It might be okay to monopopularize dandelion because there is a ton of it out there. Perhaps if this pattern of marketing is to continue, herbal product companies should concentrate on those herbs that are in abundance. Soon they are going to have to realize that the properties of goldenseal are also found in other plants because goldenseal is not going to last. And ginseng. I think monopopularizing is strictly a marketing thing. That is all it is. It's all about marketing. And ignorance. People learn about one plant at a time, basically, so they want to learn about the ones everybody talks about.

Bioregional consciousness is where it's at. I keep trying to tell people, "Learn the plants in your backyard." We've forgotten so many of them; we've forgotten all the local plants. Herbal education in the mainstream is limited to those major plants. We need to get teachers in schools to start teaching about local plants. We are all reaching into the Ohio River Valley bioregion and taking the black cohosh, the blue cohosh, the goldenseal, the ginseng, the slippery elm. The same properties are found in plants in every bioregion, but we just don't know the plants well enough. I try to tell my students that we have thirty basic herbs we teach at the school, but we are not teaching about the plant. We are teaching about the plant *properties*. Then, the student can go out into his or her local area and find plants in the area that have the same properties and start using them. We are going to have to start small to educate the local people to use those instead of the stuff that's popularly marketed in the mainstream press.

It starts small and it starts slow, but it is happening. The cottage-industry herbal businesses are making a resurgence. There are some strong examples of this kind of local business and local teaching going on in pockets across the country. You set up examples. You write about them. You present these examples to people, and you show how it can be done. There's a lot going on in the Southwest right now. I think Boulder has quite a community going up there, and I see Montana is going to have a strong community; there are a few people up there. I just keep having great faith. I keep running my school. I get thirty students a year through here, very powerful personalities, very energetic, and very enthused, and they go out. They leave the seedpod. I don't know where they're going, but I now they are landing somewhere. I know the seeds have taken root somewhere. I just keep pumping what I can in here, and I know those seeds are being carried everywhere. I know that there is stuff happening out there. It just happens kind of silently. I know a lot of my students in school are doing wonderful work out there right now, and here, there, and everywhere. So I think it's really happening. I have great confidence in the strength of heart-centered herbalism.

Chapter 15

David Hoffmann

David Hoffmann is a medical herbalist and prominent herbal teacher with a background in environmentalism and ecology. He practiced clinical herbalism for many years in Wales and cofounded a spiritual community there in the Preseli Mountain area.

In 1987, he relocated to California and became quickly involved with the American herbal renaissance. He directed the California School of Herbal Studies for a time, and also served as the president of the American Herbalists Guild. At present, David teaches on the herbal circuit about ecology and conscious-altering plants. He has written several books among them, *The New Holistic Herbal* (1997), *Medical Herbalism* (2003), and *Herbalism: Gaia in Action in Therapeutic Herbalism* (1992), an intensive correspondence course in medical herbalism.

I spoke with David at the Green Nations Gathering on September 26, 1998.

You have got to understand that I have a very extreme take on everything that is going on. I see capitalism as the cause of all of our problems, compounded by patriarchy and the church. Coming from that perspective, I have got a lot to say.

HISTORY OF AMERICAN HERBALISM

I came over here in 1986, not because of herbs, not because of herbalism. I had a very busy practice in Wales, seeing fifty to sixty people per week. And I was getting very successful. That experience gave me a really solid basis in knowing that plants work, and not an intellectual knowing. I trust them because I have seen what they can do and know that different people respond in different ways. That didn't come from my learning; that came from my experience.

I moved to California twenty years too late, but . . . I went to, basically, the hippies' old people's home, Sonoma County. I did some work at [the] California School of Herbal Studies, and Rosemary [Gladstar] asked me to take over because that was when she was moving to Vermont. So, all of a sudden, there I was, new country, running the country's only herb school, which was a very strange experience.

In 1987, there were maybe fifteen, twenty herbalists, and we would all meet at the conferences. We were the only people at the conferences . . . and there were only two. They were basically ways of hanging out with friends, which is totally surreal for an Englishman.

ESSENCE OF AMERICAN HERBALISM

Something I deeply, deeply appreciate about American herbalism, and it is one of the most nurturing things in my life, is that all of it until five years ago when the money came in . . . All of it was an expression of people following their hearts and then healing. There were no role models. There were some books, but they were terrible. You try using John Lust and Dr. Christopher and get anywhere. They were good practitioners, but they never did a class in creative writing.

What I am about to say is not a "put-down." American herbalism made it all up. And then they talk to the next wave of herbalists as though this is the tradition. And yes, it is because Rosemary [Gladstar], Michael Tierra, and Michael Moore made it all up. Not made up the knowledge of what the herbs can do, because that is in the books, but everything else. There is no American herbal tradition that is living other than Native American.

There is a heart connection with the Earth [in American herbalism]; there is a joyful celebration of that. Herbalists don't get depressed. They get depressed about [their] lives, but you can't be pessimistically depressed and nihilistic if you know what happens when you hug a tree. U.S. herbalism . . . it was grounded. Everybody was coming from that place and it was hardly ever talked about. It was celebrated, but we didn't need to talk about it. Very, very special experience.

Rosemary Gladstar basically created all of this, not on purpose. She is a hippy fairy. She can manifest things. She can organize. Everybody listens to her because she is this hippy fairy. You don't hassle. She started the California School of Herbal Studies in the back room of her herb store. It involved three people. She started arranging what she called retreats, basically, so she could get her friends together and hang out and make potpourris and do henna. These gatherings were very much influenced by her image, which was one of love and peace and no confrontation—tune into the plants and everything is fine.

Get a group of professional herbalists in a room and ask a question. You will get at least fifteen different opinions. There is no protocol. There could have been major confrontations with no one ever talking to each other ever again. It was Rosemary's synergy that created this [community] and a sort of divine blessing of Gaia would come out. So, there is a background energy of mutual respect and not the old traditional thing of different schools breaking off and fighting each other. That has started to happen now because it has gotten so big.

It is not simply the advertising of St. John's wort. All the changes that have happened since 1985, since 1990, are a result of the wider American community starting to perceive, in little ways, that there is something going on here. These people know something. Then come the FDA people at the conferences asking, Who is leading all of this? Where is the mastermind? Which advertising agency? Herbalism is a grassroots movement. It doesn't matter how often it gets suppressed or wiped out; it comes back. What is perhaps unique about the American resurgence is that basically it was a resurgence of the hippies. All of the people from the community, most of the people who are teaching here, were flower children. I hate that expression, but you know what I mean.

[The teachers] are coming from the connection with the other. (I have to use old hippy language, because it has never been updated.) We were all doing it in the streets. There were no leaders; there mustn't be leaders. I think of that Bob Dylan line, "Don't follow leaders; watch your parkin' meters."[1] The herb revival started as total anarchic green rediscovery. If it wasn't for the fact that these plants, the devas looked after us, we would all be dead by now.

The herb world is one of the very few parts of the movement from the 1960s that is still going—more than going. It is becoming a major force. When people start looking or orienting toward the green, the green touches you. And now people buy the books or come to [conferences] or listen to the tapes. That sense of the plants talking to you really shakes stuff up. And it is cool, but you try talking to your career advisor about how to become an herbalist. It just wouldn't work. So it had to be hippies that responded because we were already taking a risk. It was an obvious risk, environmentally sound, psychedelically sound, and it smelled good. It just worked.

So, I think the herb renaissance, which is throughout the Western world, is the result of hippy herbalists: Rosemary, Christopher Hobbs, Michael Tierra (although he will never admit to being a hippy), Susun Weed, Feather [Jones].

The only one of the groups of people who were doing herbalism, who still are doing herbalism, are the Mormons. There is this long-standing strand in American herbalism which is led by dogmatic Christian fundamentalists. It plays out in some very strange ways. One of the first books to be printed in the colonies as a herbal was John Wesley's—you know Wesleyanism—Calvinistic, puritanical . . . you are born in sin and will die in sin, so join up now. There has been that basis to street folk herbalism always in America or there is the Native American approach. The implications [of the former] are based in using herbs to cleanse the body of sin. Herbalism here is really colored by techniques which do not fundamentally trust the body to be self-healing. The lower bowel cleanse empties the body of poisons. There are no poisons in the lower bowel! It is not there, but things really shift if you see the body as being sinful. Hippies don't see the body as being sinful. So, we were prime candidates for Gaia.

I think I have come to see it more clearly because I am a tourist. There are herbal subcultures. There are the naturopaths who are becoming increasingly herbal. No herbalists would consider them competent, but they are competent in medicine. They have the influence of making the double-blind study more important than hugging trees, which is a bad move. There is Cajun herbalism. Somewhere in Manhattan there is an Ethiopian herb shop for that community, and on the West Coast there are all of the botanicas for the Latinos, and hardly any of these subgroups talk to each other. Most of them don't even know the others exist. Herbalism is about to stop being folk herbalism and that is dangerous. The co-option is starting to happen.

[On the other hand,] because of who the new American herbalists were, basically hitchhikers, there was a real openness to finding out what was going on in England, herbally . . . what was going on in Australia, China.

INTRODUCTION OF SCIENCE INTO AMERICAN HERBALISM

Initially, American herbalism didn't have science. The herbs worked. We were all being looked after, but herbalists didn't know any science. We could have killed people, but we didn't. So, the first real change was the introduction, the recognition that there was a need to know biomedicine. That was the beginning of people being Chinese herbalists or Western herbalists. So, the innocence went, if you see what I mean. But, it was still very much holistic. I was at a meeting at Breitenbush where I remember someone mentioning a study that they had read before most people knew about studies, and they were almost crucified for talking about dead animals . . . how dare they rape our plants like this, which I totally agree with. The maturing process started and differences were less. So, there has been the rapid development of some pretty skilled practitioners.

Then Germany started influencing the FDA.[2] The science coming in has nothing to do with us. There has always been the pharmacognocists and the scientific people to whom the idea of taking a herb is just too surreal for words. They study them. The FDA didn't realize what was going to happen. After the passage of DSHEA, more people wrote letters and called Congress about the labeling laws than anything else in Congress's history, other than abortion and the Vietnam War . . . and it was all one sided.[3] So, the FDA was put in its place by the politicians, being told, you've got to look at this objectively and do your job, which is looking after the American people. So, what they did rather than working it all out is they just looked at Germany. The German doctors do two semesters of phytotherapy. They base their herbalism, as far as they can, on rational science. Then, when science doesn't go far enough, they acknowledge that there is more [i.e., folklore], which is their saving grace. There was this ready-made body of evidence from Germany that the FDA didn't have to work at. They could just bring it in. So, the wonder herbs are the ones that the Germans did research on. No one is doing research on American plants. It is not funded.

The science came in, which didn't change anything, but then the naturopaths started to find that if they referred to this material, then they had back up documentation for some of the traditional protocols. That led to Bastyr and the other naturopathic medical schools getting accredited, which now means they have to maintain levels of science that medical schools maintain. Students at naturopathic schools are having a really hard time with it, but if they want accreditation, it is the price they have got to pay.

CHANGES IN AMERICAN HERBALISM

If the media is going to look at herbs for a feature article or something, it sounds a lot better if they get the information from an ND or a pharmacologist. Journalists have no idea what the issues are. They just pick up on the science. [In this way,] the new people in herbalism don't know about hot tubs in Breitenbush. They don't know about vision quests. They don't know about any of that stuff. And they are probably not ex-hippies. What they know is St. John's wort treats depression. It does a lot more than that. This is something the teachers are talking a lot about these days. We have to deconstruct the new herbalism—not telling them they are wrong, but put the little bit of facts they know in the real context and wash away all the bias that has come in from the science . . . bias in the statistical sense.

If you came into herbalism this year, and I came into it last year, you would see me as an experienced, long-standing herbalist. There is no sense of proportion, which on one level is fun, but it's really dangerous for therapeutics. In terms of medical therapeutics, U.S., Western herbalists are really incompetent and dangerously so. In terms of knowing the herbs, and in terms of knowing the herbs in their environment, and the wildcrafting skills, and the medicine making skills, they are the best herbalists in the world. The problem is that they don't see that there is a difference between these two types of herbalism. Knowing about herbs doesn't tell you about hemorrhoids. English herbalists are really bad at field herbalism. You never want me to make a medicine for you. When I demonstrate lotion making, it ends up over the walls.

THE FDA AND THE MEDICAL ESTABLISHMENT

The FDA and the people that publish the pharmacopoeia are realizing that they have to embrace herbs, because everyone is using them. The pharmacopoeia is about to publish a herbal *PDR [Physicians'*

Desk Reference], which sounds like it would be really good.[4] What they have done is translated a German *PDR* equivalent. They are trying to waive any responsibility. It is their responsibility to write the monographs so that you've got reliable information. They are buying the German books so they don't have to do it. They just want to co-opt it so we carry on doing what we are told by them.

The American herbal pharmacopoeia and a bunch of other books that are about to come out should begin to balance that out, but there is a move started by the medical establishment that has power, as opposed to doctors and pharmacists . . . the organization. They are threatened. They are used to having their way. Unfortunately, only thirty percent of them pay dues to the AMA now [Neel and Bloom, 1996]. Three years ago it was sixty-five percent. They are losing fast. So, there is an attempt to show to their members and would-be members that they are really embracing all this positive good stuff. My interpretation is that they want it to move in the direction that says, herbs work and they are powerful medicine, so they need to be prescribed by complete medical practitioners, therefore the doctors, therefore our members.

I was speaking at an AMA conference three years ago for recently graduated MDs in the western states. It was a whole day that was supposed to be the AMA showing how they were open to alternatives, and there I was with a bunch of insurance agents. The AMA totally blew it. The students got it immediately. The only reason the medical students join the student branch of the AMA is because you get a forty percent discount on books. They never join after that, and this was an attempt to get them to join.

I realized when it was my time to go on, there was no way I could talk about herbs. In the context, nothing would work. They had no frame of reference. I had just been working on a book on information retrieval for herbal medicine. So, I did an off-the-top-of-my-head presentation about the herbal information in their library because it is a really rich source. Medline [http://medlineplus.gov/] is a really good source for herbalists. The people at the conference had never even been told about these classifications in the Dewey decimal system. The statements that came up afterward, apart from a couple about specific books, were about how angry they were that they are in debt for the rest of their lives to pay for what was supposed to be a well-rounded education in medicine, and nobody had ever told them about these mesh terms and ways of getting information out of Medline. They are really open, but they are educated to be closed.

What is happening in this country with herbal medicine and supplement medicine is that people see herbalism as herbal products. The herb industry is just a bunch of people making brown evil-tasting tinctures that somebody else uses. Herbalism is about how to use it. The idea of people asking salespeople in the herb departments in stores for medical advice on how to use herbs is dangerous; that is where most of the therapeutics go on. On that level, I totally support the FDA because people are going to get killed—people have been killed. A little bit of knowledge is an exceptionally dangerous thing. What should be happening is that anybody, any storeowner, or chain with a herb department, should pay for education for their staff. They don't. They may pay for people to come to a couple of [conferences]. These are fun and you can meet people, but there is not enough education going on here for it to be the basis for someone's knowledge or understanding.

HERB INDUSTRY

[Herbalism] has moved from the grandmother in the cottage—Rosemary [Gladstar]—to what it is now. The only reason so many little herb companies formed was that you couldn't find herbs. So, every herbalist made their own, which hasn't happened in Europe because you can always get herbs. There are a few of those small companies, which have become incredibly successful. You can do it right. Herb Pharm, Herbalist & Alchemist are really herbally ethical and pharmacognostically ethical. They really make sure their stuff is right. They are not ripping anybody off. They really are holding a vision of being the suppliers to the green healers. That is not what many of the new herb corporations are doing. They are providing products for the green health industry that aren't going to work.

If you are a herb company competing with other herb companies, you don't advertise herbs. You advertise why your processes are better than everybody else's. That has generated a wave of people coming to herbalism who believe the advertising and who really want to know what is the best way of taking herbs. The best way to do it is to lay on the grass and chew. The new, corporate herb industry is a bunch of bloodsuckers. I can give you many, many examples of products that are out there purely for profit. They are not good products. People have been lied to, ripped off, and, worst of all, have had their hopes and dreams blown. Somebody with severe problems, cancer or whatever, may start taking herbs. It is going to be them making a step toward self-healing. That, then, gets turned into a sales event that then rapes them. You can't have healing in the marketplace. You can have pathology treatments, but profit and what that implies are anathema to spiritual healing, the transformation from within. They are mutually exclusive.

HERBS VERSUS HERBALISM AND THE HERB INDUSTRY'S POSITION ON THIS ISSUE

What has happened in the herb industry is they have got everybody not thinking . . . just knowing that herbalism is about herbs. It is not. It is about our relationship with the planet via herbs. What the media don't get is the difference between the herb and the protocols of the herbalist, the experience, and just knowing the issues. The AMA are feeding on that because if it is just the herb, they can show quite easily that with science they know more about them than we do. The herb industry has been promoting herbs, not herbalism. People think they can go and buy a bottle of something and it will be a magic bullet. It is the herb industry's fault.

My positive motivation in herbalism is working with the plants for the alleviation of human suffering. Not, "I am promoting herbs." One-quarter of all prescriptions are still made from herbs. Three-quarters started in herbs. I don't see any problem with another way of using herbs. What I see as a problem is pharmaceutical use of herbs closing the doors for people to get a relationship with Gaia going through their use of herbs. They become substances, and sometimes substances are appropriate, but they are no longer herbs. I don't know anybody who has ever had heart communication with a standardized tablet or extract. That is the issue for me.

SPEAKING THE LANGUAGE OF SCIENCE

I use scientific language so the medical establishment can hear me. The only way people can ever hear anyone else is if the person is expressing the issue in their language. We have to communicate with other interest groups. I don't think any of us are doing it because we want to turn them onto herbs. We are doing it because they are using herbs, and, if something goes wrong, we're going to get blamed because we are the herbalists. The medical establishment gets so anal-retentive about words. Use the word *tonic* with pharmacologists and they will leave the room. Use the phrases *immunomodulation* and *psychoneuroimmunology* with a few buzz *saponins* in there, which is one way of sort of distorting tonic, then they will hear you.

I am into communicating with the medical science community for political reasons, survival reasons. I think we need another five or ten years of nonoppression so that the alternative health care system can be in place for when the other system collapses. One way to keep the AMA and the FDA away from herbalists so they don't do a Flexnor again is to talk to them in their language so they think they are leading it.[5] Now, if you start talking about saponins, and they know we are just herbalists, they know we got the information from them. The doctors can really hear what we are saying. This is just a way of getting them to back off.

According to Trotsky, in order to generate revolutionary change, you help the enemy go to the extreme of what they are trying to do because the seeds of the destruction of the system are built into capitalism. The seeds of the destruction of patriarchal medicine are in the edifice and we can't break it down. It has been tried. It can't be done. The edifice is strong. It is crumbling from the inside. Speaking the language of science is throwing them a bone so we don't become the next "red under the bed" scare.

Every time in English history that there has been a social upheaval, herbalism has had a major revival. Culpeper is the best example of that. His place was burnt to the ground three times by the pharmacists. This could happen again. It wouldn't take that form. They would just bankrupt us all. It would be litigation, the modern version of burning at the stake.

In theory, you still could be put in prison for practicing medicine. I had a line in an article that I keep finding on the Internet. "I can get a license for a gun, but I can't get a license to practice healing . . . only in America." Nowhere else in the world is it like this. The best herbalism in the worst environment; it is a really good recipe for rapid growth and change. It is going to go in the right direction because at the heart of American herbalism is what Rosemary [Gladstar] is doing and what Pam [Montgomery] is doing. We actually can't go wrong as a movement because we don't know where we are going.

We should be giving our stuff away freely, our knowledge, our experience. The idea of trade secrets in herbalism, or secret formulas is a death net. It really is. That is what killed the Eclectics—all of the infighting. We need to empower ourselves with relevant herbal information from where we live, and then stop being as naive as we are as a community. I don't mean everyone has got to learn to build barricades and make Molotov cocktails. Every herbalist has got to know where the local child guidance center, or social services department is, who all the local practitioners are, who the local environmental groups are because they know where the herbs are. I am working on this booklet, which is going to be adjustable for any bioregion. Part of the booklet is going to be a template that will help people pull facts into it. If you know your flora and you know how to find out what the actions of the plants are, you have got an instant materia medica that can be translated in terms of action rather than specific plants. The person could use this template to find out which local plants are used for stomach ulcers.

EXPANSION AND POTENTIAL OF AMERICAN HERBALISM

Our gardens and our little medicine-making space in the kitchen are so nice. It is like this nurturing little bubble of green. There are lots of herbalists who are shocked that there is anything outside of that bubble. In the ten years—well, I have been coming to this for nine years, I have been watching a change from bubble people to a more socially relevant group of people. They have expanded their bubble; they haven't left it. Miraculously it has worked. We are being looked after by the green. You are going to come across lots of people who realized they had to become herbalists because there is something inside of them drawing them to herbs. What is that? If it is intuition, what is it that we are intuiting? It is happening everywhere to people who are really surprised by it. It is one of the main ways in which the life raft has been thrown to us from the elementals. I am not saying that herbalism is going to save us. I don't think it is going to save us. It is a part of living lightly on the Earth and alleviating suffering at the same time.

One of my agendas for traveling so much is to help with input into local town herbalists and herb gardens. I have never been anywhere in this country that hasn't had an herbalists and an herb garden. They may not even know each other. There may be two herbalists living on adjacent streets and who do not know each other. But what happens when the petrochemicals run out and there is no distribution system, when there are no raw materials for making drugs? It will be messy; it could be medically scary. Apart from acute infections, we know what to do . . . and there is now one of us everywhere. I want there to be lots of us everywhere, so that when the shit hits the fan, people don't have to suffer as much until whatever comes after it manifests. When the pharmacists can't fill the doctors prescriptions, they are going to remember their training, and they are going to remember that the anti-inflammatories come from plants. We will be there. We can tell them which ones they are.

PLACE OF SCIENCE WITHIN THE AMERICAN HERBAL COMMUNITY

When nonscientists learn science, it is really bad, really bad. The herb world embarrasses itself. The pseudoscience is just gross. The strength of herbalism is in folk wisdom and the fact that the herbs work. The only people that should get into the science are people who know the science enough so they can step out of the science and ignore it. If you start from the place of really being embraced by Gaia, and you are attracted to science, then it is worth doing the two.

Science does not make herbalism better. It may make safer practitioners. It may avoid some really dumb things, but it doesn't improve on the herbs. The technology that is driven by science—standardization, freeze-drying, all that stuff—is one of the things that has gotten so much attention. It can sound new and flashy and breakthrough, but it uniformly ruins herbs. The only time I will think of using the new technology is to make the herb which is too unpalatable, palatable.

Saw palmetto berry extract in capsules. If you have ever chewed on a saw palmetto berry, it is obvious why we put it in capsules. Or standardization for herbs where the effective dose is unbelievably high, for example, milk thistle seeds.

In England and Germany, you get competent medical people who know herbs. I really see the relevance and the value for someone to learn to be an herbal practitioner in Western times, but not all herbalists have to become clinical herbalists. Some herbalists should never be around other human beings. They are plant people. I can kill dandelion when I try to grow it. I can't grow plants. They know I think I can't do it, so they don't like me. I am a wild plant person. People can be herbalists who never look at hemorrhoids in their life. People can be herbalists who have never grown dandelion in their life. It is very diverse and therefore ecologically very stable.

PRACTICING HOLISTIC HERBALISM

If somebody has an illness caused by their work ... some industrial illness where the main causal issue is the work environment or the stress or whatever, and they come to me as a holistic healer who has studied the Hippocratic oath, I would interpret it as my responsibility to tell them to not go back to that work, not to do what most herbalists and most MDs do—patch them up so they can go back into the trenches to be shot again. It is a real difficulty. I choose to be extreme. It is not difficult for me to have an extreme stance. I will tell people like that, I am the wrong practitioner. I am comfortable sticking out like a sore thumb. Most herbalists aren't. They want to help. They want to be there for people. That is probably not the best attitude for clinical practice. You need to be a bit more objective. In practice, we have to find a way of helping the patient reach the point where they see that going back to their job is inappropriate. Otherwise they haven't got it; they will just be doing what we say. And how do we know what is appropriate for somebody else's life?

I much prefer to give people packets of seeds than bottles of tinctures because then there may be the possibility that they get the intellectual realization that this is a plant. It is the environment, not the brown liquid. The herb industry could do something about this. The bottle of tincture could come with a packet of seeds. It would be really good advertising.

We can use the protocols, either the counseling or the herbs, as a way of helping a person come to an issue that might bring up a realization. I can rant on for hours about green connection. I never bring it up in medical practice. It is inappropriate, but the first hint from a patient of anything having to do with awareness of a plant or the environmental connection . . . I use that as a green light to then start helping them to bring up the issues. I wouldn't bring the issue up, but I can use herbs and counseling techniques to maybe change their minds a little bit, move their perception a little way so they might ask questions. You don't even have to ask the right questions because the green gets into us. Just being willing to question allows it to come in. There is a line, "Take one step toward God and God will take a thousand steps toward you."[6] We are educators. That is what it comes down to. To teach means to draw out, not to lay on.

LICENSURE

I support peer review licensure. Most herbalists don't like it. It can seem to read like the regulations of other fields. It seems like it is the government licensing you. It is not. The state acknowledges the field as having skills that need licensing and then a peer group from that field does the licensure. The reason we need to do it is for the safety of patients. It is not to further our egos or money. There are some really dangerous ideas out there—bad medicine, herbs used inappropriately. We herbalists shouldn't be doing that. One way of protecting the general public, who are coming from herbal drought, is to police ourselves.

The patient needs to know that the herbalist is competent in his or her branch of medicine. Knowing the herbs is not enough. Usually with licensure is the legal right to diagnose. Herbalists at the moment shouldn't diagnose. They can't diagnose unless they have done something specific to learn it—people who have paid their dues doing other things [i.e., TCM, Ayurveda]. With the general education in American herbalism, there is nothing competent about diagnosis.

My reason for supporting licensure is to try to minimize suffering. That is the reason, not any legalistic reason. The herbalists who are most opposed to licensure are the herbalists who don't know any medicine. It gets very difficult in group meetings because it ends up sounding like the medically trained herbalists are being elitist about the other ones.

Because I never did a PhD, for a very long time I felt very strange when I was walking through a university campus. Why didn't I do this? These people know something. I dropped out too soon. That is the conditioning in me from the old days. I got rid of it, but I see that same thing going on in the herb world. People who have incredible wisdom and knowledge and such inherent self-worth feel that they are not competent enough because they haven't done this other thing and then reject it, exclude it. It doesn't have to be like that.

There has to be a recognition that there are health healing herbal modalities that don't use science. When licensure comes, we herbalists who are in charge of the writing of it have got to make sure that the wording is such that there is no hidden iron fist to get the folk herbalists, which is what happened. . . . The Eclectics were invited into the new system; the folk herbalists weren't. There needs to be licensure for medical herbalists who say they do Western herbal medicine because that implies they are competent in Western medicine. The TCM herbalists can go for TCM licensure, but that doesn't cover you for Western techniques. So does Ayurveda. I don't think there can be one license. It is going to be really strange and messy. It could be as disorganized as the naturopathic degree. If you get a naturopathic degree, what do you do? Are you a naturopathic homeopath? It is very confusing for everybody. If the herbalists tried to do that, we would make it even messier.

When we formed the American Herbalists Guild, there were twelve of us who did it. The first four years at every conference we tried to get our membership thing together and what the criteria were. We were trying to embrace, in a meaningful way, Native American approaches, medicine wheel approaches, Wise Woman approaches, channeling, drug chemistry, pharmacology, TCM—all in the same thing. We finally got over that by ignoring the issue. We just acknowledge that people have competent skills in their area of practice, without saying what that is. How do you quantify spirit? But then how much of that is actually going on in overtly medical situations anyway. What the licensing would be for would be those herbalists who are in the yellow pages as clinicians, not those herbalists who take people on vision quests or do sweat lodges.

The licensure is there as a safety net for people. Those people that get licensed then have a major educational job to do. We would have to make really clear that we are just putting it into the culturally acceptable form of health care, this wild and crazy stuff, and the old ways. That is the point where I don't trust the American herb community. I don't think, on the whole, they are articulate enough or even perceptive enough with what their heart is doing to get that message over.

The people in the herb world who could get their positions over are naturopaths. The people like Michael Murray, who is a really good pharmacological, botanical medicine person could do it. He is not a herbalist, but he has a really good media presence. It is a challenge. It is not the doctor-herbalist challenge. It is the new form of herbalist-general public challenge. We are going to lose an opportunity for creating more doorways for the green to manifest and the healing to happen from the inside out. We cannot change the monolith; the weeds will grind it down bit by bit. We can speed that up by turning people onto their roots; herbal products don't. They just help people lose weight in their bank account. It is going to be interesting to see what happens.

INTELLECTUAL PROPERTY RIGHTS

I think that [concepts of intellectual property rights] are the most evil incursion of the capitalist worldview that I have come across. I am communal, period. All of my books are considered to be public domain. Michael Moore does that. All of his stuff is there.

There are intellectual property rights, but that is just a way of absolving the conscience of ethical capitalists. They are still ripping off these cultures. What there needs to be is an ethical respect for our brothers and sisters who wear different clothes. That is not what is going on here. The pharmaceutical industry needs raw resources. They are found by their field workers, the ethnobotanists. The ethnobotanists tend to have a much better heart sense of responsibility than the drug companies. So, yes, they started listening to people from these countries talking about rights, and then they think they have done something about it by actually acknowledging intellectual property rights. The English expression is "pissing in the wind." The idea of coming up with a conservation plan for a forest to conserve it for future generations—we are just acting as the bankers for the planet rapers. We need to stop them raping, not train them to rape nicer. What do you do about profits being made from insights coming from cultures where they are starving? There has to be some sort of equity.

As soon as money is brought in, spirit goes out. I don't think we can reimburse rainforest communities for the destruction of their [lives], liberty, heart, future . . . and they are all going to die of diabetes from the food really soon. We just need to acknowledge that our culture rapes. It rapes everywhere it goes. The only reason New England didn't turn into desolation was the people who raped, then went west. The trees here have regrown. It is wonderful. This is probably third growth, but at least it is growth.

The pharmaceutical companies want the phytochemical knowledge. That may be the saving grace because that is not going to help them much. . . . Take for instance the work on St. John's wort, where they were saying the active ingredient was hypericin, and it is not. There were companies who started making products with extra hypericin in it with all of this evidence for doing it. They had fallen for it. I don't think the pharmaceutical industry will ever be open to what herbs do because you cannot patent a plant.

The pharmaceutical companies are buying out the natural product companies. They are bioprospecting my experience. What I find sad, but I can understand it, is that many of my colleagues actually fall for it and become consultants and help them. I can understand it because our community is so naive in the ways of the world. You can't trust bankers when it comes to the Earth and Gaia. It is one of the reasons I get very political at these events—not because I am trying to teach something political; I just want to raise politics. It used to be that people would just walk out of the room. They are starting to get it now. I don't want people to agree with my politics. We just need to have a political differential diagnosis about herbs and things. Gaia does not have intellectual property. It is communal. We are being talked to.

Chapter 16

Ellen Evert Hopman

Ellen Evert Hopman is an herbalist and lay homeopath who holds an MEd in mental health counseling. For nine years, she was vice president of the Henge of Keltria, an international druid fellowship, and is currently Keltria's international coordinator.

Ellen has been a teacher of herbalism since 1983 and of Druidism since 1990. She also coleads tours to celtic and neolithic sites in Europe periodically. While Ellen has written a number of books and articles and has released audio and videotapes on the topics of herbalism and paganism, her co-authored volume of interviews, *People of the Earth: The New Pagans Speak Out* (1996) had an enormous impact on me and served as a model as I put this project together. Ellen was also featured in "The Unexplained," an A&E television documentary about Druids.

Ellen and I discussed the topic of herbal education over the phone on August 16, 1999.

COMING TO AND LEARNING ABOUT HERBALISM

In common with the rest of my life, I am a mystic. Everything happens to me mystically. Nothing happens to me in a logical, rational way. To make a long story short. I was in Italy working on my master's in art history at the time. This is in the very early 1980s. I traveled to Assisi to look at the frescos by Giotto in the cathedral there. I went to a place called San Damiano which is a convent that Saint Francis built for Saint Clare. While I was there, I met a monk and I asked him in Italian—I said I have been to the cathedral. I have seen the frescos. It is all very elegant, but I want to know more about how Saint Francis actually lived because I know this doesn't bear much resemblance to who he was. The monk said go to San Masseo. I said what's that? And he said don't ask any questions. Just go there. I did. I walked down the road, and there was this little wooden sign pointing off through the bushes and there was a little mud trail. I walked down the trail, and, at the end, there were all of these people just sort of hanging out on the grass. They said, oh, have you come here to live? I said, huh? They said, you, yeah, you can stay here. I said, oh, okay. What is this place? They said it was a Franciscan community.

I stayed for four days, and then I went back to Rome, packed up all of my stuff, put it in storage, and went back and stayed for a few weeks. It was only a few weeks, but it was momentous. The routine was Mass three times a day—first thing at dawn, in the middle of the day, and then in the evening. We fasted Wednesday afternoon, and Sunday all day we fasted.

One Sunday, I was walking up into the mountains where Saint Francis used to go to meditate. A storm came up, and there was thunder and lightning and snow. It was October, but I was up above the tree line. The only living thing up there was a tiny pine tree. I curled myself around the tree because I was frightened. It was stupid, but that is what I did. It was just me and the tree up there. When the storm passed, I came down the hill, singing at the top of my lungs. I remember looking at my shoulders. There was actually snow caked on them.

I came down the hill and walked all the way back to San Masseo. There was a little Romanesque chapel there that Saint Francis had actually worked on, that he had repaired himself. I went in there. It was pre-Gothic. There were no windows. It was very, very dark in there. In the darkness, I sat down, and I heard this voice. The voice said that everything that you have been doing up until now has been for status, intellect, and to please your parents. You are supposed to be working with plants.

I had never thought of that before, but it made total sense. I had a very strong heart connection to plants, the green world. The first thing I did when the storm came up was to go to a tree. That was my instinctive reaction. I was passionately in love with the wild flowers and the olive trees. To me, that was life, and all the stuff that I was doing in school was dead. It was art history. It was all the past, and it was dead, and it was books, and it was buildings. I knew intuitively that the voice was correct. All of a sudden, I started visualizing the implications of that. I visualized giving up my degree. I had a house at the time. I visualized giving up the house. I started mentally throwing out everything that I had and thinking, *God, I have to go get trained to work with plants.* That is exactly what I did. I went back to the States and sold everything. I found a teacher, apprenticed for five months in New York City.

While I was in New York, I took classes, but we also had clients who came in, and I sat in on that. I also went out to buy the herbs with the teacher. We went to Chinatown and checked out all of the herb shops. We learned how to tell the difference between good astragalus and bad astragalus. There was more to it than just sitting in the classroom. That was in 1983. I started teaching immediately and I have been doing it ever since. Then I started writing books. That is how it happened.

I don't know if it was just my intuition that directed me to my work with plants. I have a feeling that there are certain spirits that work with me. Sometimes they have to be very obvious because I am sort of thickheaded.

TEACHING HERBALISM

Six months out of the year I teach students of herbalism. I have an intensive that starts in October, and it goes through April. They come to my house every week for six months. I am completely absorbed with these people for half the year. At the same time, I am being invited to teach all over the country and internationally. I have done workshops in Scotland and Ireland and Canada as well, both in herbalism and paganism.

In my six-month intensive, I have the students start out with the theoretical knowledge, which is something that I find most students really lack. I find a lot of people who call themselves herbalists who know very little, and who have been told that their intuition is everything. When I listen to them, I just cringe. Some of them can be very good. Over time, I think, if you make enough mistakes, you learn. How did Paracelsus put it? I think it was Paracelsus who said that intuition without experience is blind, and experience without intuition is also blind, although he put it a lot more elegantly. You really need both.

I start my students out with a very good theoretical underpinning. I teach them the basics of Chinese five-element theory. I teach them facial diagnosis, the doctrine of signatures. Then they get the idea of flower essence counseling. I teach them the idea that the mind has a profound effect on healing. They also learn the herbs. I have over four hundred pages of handouts that talk about materia medica and specific conditions.

Once they have a strong theoretical underpinning, they have to learn to put together formulas. There is another whole science to that. I was taught to always have a minimum of three herbs, which is something that most people don't get to do. You need an herb to build the system, an herb to cleanse the system, and then an herb that is tonic to the system, or whatever organ you are dealing with so that there is a movement. Things need to be moving through, building up, and clearing out. You don't just take one herb and throw it at the body for a long time. Again, that is the kind of thing where people need training, and I see that people really don't have that.

Once my students have all that, we start doing the actual case histories where we bring people into the class. I am a strong believer in the group mind. I think that, especially in the beginning, it is important to have everybody take the case together because one person will pick up more on the psychological aspect while another person will pick up more on what is going on physically. I always say six heads are better than one [laughs]. So they take the case, and I divide them up into groups. Each group goes and sits in a different part of the house. They figure out how to take care of the case. They come back and report to the group. Every client that comes to the class gets a flower essence to deal with the emotional side of what is going on. They also get an herbal formula to deal with the physical aspects of what is going on, although we also understand that herbs have psychic aspects as well. We pick the herbs with that in mind. Then they also get general recommendations for their life, like they are not getting enough exercise or they are very depressed in their career. They need a career change. Or they need to do something about their marriage. Whatever, it is something about their lifestyle. When the students walk out of here, they are armed with a huge amount of information and tools.

VALUES SHE COMMUNICATES TO HER HERBAL STUDENTS

One of the things I tell people is that if you are going to be able to see what is going on with another person, you have to be pretty balanced yourself. The other thing I tell my students is not to have any preconceptions. That is something that bothers me about these different "schools" of herbalism. Each school has its own theoretical preconception. When a client walks through the door, they try to fit the person into their theoretical preconception. I heard one herbalist say it is all about nourishing. Just nourish, nourish, nourish. So the students could walk away thinking whoever comes in the door, that is what they are going to do—nourish, nourish, nourish. That particular school of herbalism also has a complete fear of people needing to cleanse in any way. When I was in New York with my teacher, we had people coming in who were blue-collar workers of a particular ethnic background who had very bad diets. They were eating tons of fried foods. They were grossly overweight. They had very bad skin. It was perfectly obvious that what they needed to do was to radically change their diet and they needed to start with some kind of a cleansing program. Just briefly, not make a lifestyle out of it. Just to kick it all off with maybe a few days of juice fasting or raw vegetables or something. Sometimes that is necessary.

You never know who is going to walk through the door. You might have someone where ninety-nine percent of what is going on is emotional. You might have someone else who has had a blunt trauma or something injured him or her and it is an old injury. That has to be dealt with on a physical level.

Another thing that I tell them is that case taking is mostly listening. Let the client tell you. In the first five minutes you should really know if you are dealing with something strictly emotional or whether you are dealing with something strictly physical, or what combination of the two.

WESTERN SCIENTIFIC TRAINING FOR HERBALISTS

There are certain things you need to know. You need to know where your liver is, for instance. If someone says, "My stomach hurts," you need to say, "What do you mean? Where?" And then they point to their spleen. There are certain absolutely, rudimentary basic things that I think everybody should know. I know some people really stress this. I use pictures of the body to help students get a rough idea of the body. I do recommend *The Anatomy Coloring Book* [Capit and

Elson, 2001]. You do have to have a sense of some biomedical science.

ROLE OF THE CONTEMPORARY HERBALIST

The other thing is that I tell my students over and over again that they are not doctors. It is not their job to diagnose. We do some facial diagnosis, but that is just to give us an idea of whether we are working with the water element or the fire element, or earth, which corresponds to certain organs in the body, but in a general sort of way. What I tell my students over and over again is when you work with someone, use a doctor to do the diagnosis. They have the advantage of the machinery. They can do blood analysis. They can do all of the testing. The client then has to come and tell you what is going on. It gets really interesting when the doctors can't tell you what is going on. That is fun because then you are free to use everything that you know. So the client then comes in and says it is this and this. Then you say, based on what I know or based on the literature, these are some of the herbal options. Of course you are paying attention to the [client's] diet and to [his or her] emotions. But my students are not doctors. They are not pretending to be doctors. It is not their job to diagnose and prescribe. It is their job to share their knowledge and give the person a different approach to [his or her] condition.

If a student wants to be a doctor, I tell [him or her] to take premed classes and then go get your naturopathic degree. Two of my students have graduated, and one is right now in premed firmly on the track to doing this. That is what I counsel them to do. If they really want to be a doctor, then they should be a naturopath. But an herbalist is something different. We are not doctors.

An herbalist is somebody who works on a much more basic level. They are much closer to nature, I think, in some ways. The naturopaths are getting awfully fancy these days. It is kind of distressing actually. Instead of using whole herbs, they are into using standardized extracts. The herbalist is a lot closer to the Earth and works on a much more basic level. Probably if you are a long-term herbalist, it is because you have a very strong spiritual relationship with the herbs and the trees. They are like your family or you wouldn't keep doing it. That is your world. You live in the green world.

MEDICAL SCIENCE AND ITS INTERFACE WITH HERBALISM

I had gone to a big herbal conference, and I heard a naturopath talking about how saw palmetto didn't work and how the only way it would work was if you bought his standardized extract. So, I wrote a letter to the AHG (American Herbalists Guild), which they published in their newsletter. I got a lot of positive feedback. I said well, that is interesting considering that Native Americans have been using saw palmetto for thousands of years. Boy, those Native Americans were pretty stupid. They just kept using it, and it wasn't working. Isn't that dumb [laughs]. People wrote back, and they thanked me for saying that because you have these—they are almost like hucksters. They get up in front of the audience, and they show you all these things—they show you these fancy charts. Wow, you know, I got proof here. Nothing is going to work unless you take my extract. They give you all the charts and figures. I am sitting in the back of the room fuming. People have kept this information for thousands of years, and humans don't keep stuff like that if it doesn't work. So, anyway . . .

This for that. It is like any science. Medical science is getting the thinnest, top layer of a vast universe, and they don't even know what they are missing. That is why herbalists are important. We need people in this world who are really involved with the plants, who focus on the plants.

My father died of cancer in 1988. The whole time he was sick . . . he was sick for seven years, I was getting him all these books about macrobiotic diets for him to go on, just for a short term, not to make a whole lifestyle about it. I was talking to him about garlic because I knew there had been all of these studies done on it. His doctor would ridicule everything that I was saying. One day, I handed my father an actual scientific study about garlic that he showed to the oncologist. My father said the doctor threw it on the floor and started screaming, "Why do laypeople always want to get involved?" That is what the oncologist said. My father died in 1988, and two years later the American Cancer Society came out with their first pronouncement that well, gee, yes, diet might have something to do with cancer, and you should probably increase your fiber and lower your fat [Willett et al., 1990]. This was what I was trying to tell my father all along, but the oncologist kept saying there is no proof of this, making my father feel that I was some kind of a total idiot. That was one of the most difficult... here is my father dying in front of my eyes, and from the moment he was first diagnosed, there was nothing that I could do. That is a whole issue in itself. The whole rescue thing where you want to rescue somebody that you just can't. I guess it wasn't his time. He was born in the wrong generation. He was not in the right mind-set to receive the information.

That is the thing. It is a lack of respect for tradition and history, lack of respect for indigenous people and indigenous cultures, a lack of understanding of indigenous people. These people survived for millions of years with very little to work with, so what they *do* have to work with is very useful. There is a complete lack of understanding and respect for indigenous wisdom, although I think slowly people are starting to realize this. It is because we are losing the rain forest; we are losing the indigenous people. That is starting to seep into people's awareness. Both as a pagan and as an herbalist, I have a tremendous respect for the ancestors, and everything that they have handed down to us. I am very interested in keeping that stuff alive.

ESSENCE OF HERBALISM AND THE NECESSARY CONNECTION TO THE GREEN WORLD

Herbalism is about taking care of yourself and your family. That is what it is all about. Otherwise, why do it? If you are just going to have a mystical relationship with plants, then that is not herbalism. That is spirit medicine or something completely different. Herbalism is learning how to use plants for medicine, for food, for crafts, having a deep relationship with nature, knowing what to gather at what season of the year. If you are an herbalist, you know, you have your own wheel of the year. There is a time when you get the mugwort. There is a time when you get the horse chestnuts; there is a time when you get the walnut hulls. There is time when you pick the hawthorn berries; there is the time when you get the roots; there is the time when you get the flowers. You are living with the green world in a very intimate relationship. Those are your buddies. Those are your friends. Those are your counselors. You are using that knowledge for your own health and for other people. If you are not doing that, then you are not an herbalist. You are doing something else. Maybe you are being a mystic, you are practicing a nature religion, or being a nature mystic, but that is not herbalism.

One reason why I have such a strong theoretical bias is because my teacher was in New York City. He did take us out on herb walks, but that was the extent of it. We got to go out on herb walks in suburban areas. But I live in the country now. I live on twenty-five aces of land. I live next to a wilderness area. I am surrounded by green stuff, and it is very much an integral part of my life.

What I do is after students are done with my class, when they ask, well, what should I do next, then I recommend they go to somebody who has an herb garden, a large garden situation, and study with that person. Then they will get the other half of it. You need both.

What has happened with herbalism is that it is being co-opted slowly but surely. What has happened is more and more people are using the herbs, and the pharmaceutical companies have noticed that it has become—at one point it was a two-billion-dollar-a-year business. I am sure it is much more than that now. That was many years ago.[1] In this country, money talks. Once they saw that, then they grabbed onto it. Now you see the ads on television for ginkgo and St. John's wort. It is kind of mind-blowing. Twenty years ago there would have been no way to imagine that. The mainstream media is grabbing onto herbalism because it is a moneymaker.

LICENSURE

I think licensure is completely unnecessary and probably dangerous in the long run. In England, there is no licensure, but herbalists are respected just because they have always been a part of the culture. They are so well respected that they have hospital privileges. They can hang out a shingle and advertise that they are an herbalist. There is absolutely no evidence that masses of people are being poisoned or dying or getting in trouble because they are going to herbalists.

In Europe, they just have a lot more respect for the herbs. You can walk into any pharmacy and you will find homeopathic remedies and herbal remedies right there on the shelf. Only recently [in the United States] you have started to be able to walk into a [drugstore], and they will have one shelf with the encapsulated ginkgo and St. John's wort, but that has only happened in the last five years. In Europe, they have always had that because they never lost respect for the herbs. Their grandmothers did it, and their great-grandmothers did it, and their mothers did it. Everybody did it. You just knew that chamomile worked, and you knew that lavender worked. Everybody just took it for granted because they didn't get co-opted by the huge commercial pharmaceutical interests the way we have with all of our advertising and mass marketing. We have really been sold a tremendous bill of goods in this country and we are suffering for it. People can't afford the health care. That is another reason why people want to learn herbalism. You can treat yourself.

Once you have a licensure system, then people start suing you. That is what will happen in this country if they do that. If people really want a license, they should be a naturopath. That is the way I look at it. This is something different. This is an earthy pursuit. It is very creative. It is an art form. Everybody does it differently. It is very fluid. The way I like to approach it is half intuition and half book knowledge. I think you need both. You can't have just one without the other. If you do, you are a fool. Everybody has to come up with his or her own mixture of that and how to handle it. But if someone wants a license they should go be a naturopathic doctor.

Chapter 17

Steven Horne

When I spoke with him in 1999, Steven Horne was the current president and executive director of the American Herbalists Guild. Currently, he is a clinical herbalist, teacher, and the founder of ABC Herbs, an herbal product company and wellness center based in Utah. Steven is also president of Kether-One, an educational materials supplier that creates training manuals and third-party literature concerning natural healing techniques. He has authored many books including *The ABC Herbal: A Simplified Guide to Natural Health Care for Children* (1995).

I spoke with Steven over the phone on August 27, 1999.

PATH TO HERBALISM

I was working on actually getting a merit badge for Boy Scouts when I was a teen. I started studying wildflowers. I had a guide that talked about the Native American uses for native plants. I was on a camping trip when I was fifteen years old. I had just eaten stinging nettle. I cooked it up and ate it. That was a monumental experience for a fifteen-year-old. I was sitting on a rock looking at the different plants that I had been learning, thinking, *I use that one for dye and I use that one for fishing line.* As I was thinking about that, I had this little thought pop into my head. *If you were a loving god, wouldn't you put everything, naturally, on the face of the Earth to take for your children?* I thought about that. It was an interesting idea. I got the idea that there were plants out there to heal our diseases. I got fascinated with it and did it as a hobby for outdoor-survival type things for a number of years.

I also had quite a few health problems myself and so I was searching for answers. It occurred to me that maybe—this was several years later—maybe these herbs would actually work better than the drugs. I started using them. One thing led to another, and I just kind of got into herbalism.

NATURE OF HERBALISM

When I first started using herbs I was using them in a medical model of thinking, for example, for symptom suppression. Then I began to learn a different way of thinking about health. I started to get dependable results. Herbalism is different from the medical model of health. The medical model of health basically works on taking a particular factor—let's say high blood pressure. We give you a drug to lower blood pressure. Or if you have high cholesterol, we give you a drug to lower cholesterol. You have runny nose, we give you a drug to dry up the runny nose. That is not the model in which herbs work. Herbs work on a constitutional model. That is, in herbalism when you are really effective, you are treating the client in a very holistic manner. Right now herbs are becoming very popular, but for the most part the popularity is, like—St. John's wort is for depression; saw palmetto is for prostate problems. That is an oversimplification of what those herbs do. Those herbs work constitutionally on the entire body. When you understand that there are constitutional effects, you get far better results than if you do otherwise. There are a whole bunch of herbal antidepressants; each has a little different profile of the kind of depression, the kind of person, it works on. When you understand that, you start to get dependable results. Herbalism is its own tradition.

In my opinion, I always think of herbalism as the medicine of the people. I think herbalism should be our first line of medical defense. We should be using herbs to treat common family ailments and then, when . . . I have a book written called the *ABC Herbal*. In it, I support the philosophy that modern medicine should be the alternative. You try something simple first, unless it is a real serious thing. If that doesn't work you go to the bigger gun. I think we should try simple, mild things first unless it is a life-threatening situation.

HERBAL SELF-CARE

I support average people self-medicating themselves with herbs. I have seen people doing that for years and while I see them sometimes being really stupid about it, I don't see any serious harm. Usually, it is just they don't get good results. Maybe they have exacerbated some symptoms, but it isn't a life-threatening thing. One of the things that I tell people about the safety of these remedies is I have seen people who are taking like thirty different herbs and nutritional supplements or forty different herbs or nutritional supplements every day. I knew one lady who took one hundred twenty herbs and nutritional supplements every day. I don't know how these people afford it, but . . . they were spending hundreds of dollars on herbs. I look at it, and they have been doing this sometimes for months. One lady had been doing this for years. And I thought, *if you went to the drugstore and picked thirty over-the-counter medications at random and took them for a month, I am sure you would have some pretty bad side effects.* But these people didn't seem to be suffering any really serious harm from it, probably because they were taking a whole bunch of herbs that neutralize each other.

I think there is a place for self-medication. Then, there is also a place for trained professional herbalists. Then, there is also a place for medical doctors in the health care system, but I would like to see everybody in the country have enough training to treat common ailments with simple home remedies. It is not that hard. That is why one of my things is I teach people how to do that.

In terms of some of the controversial herbs like ephedra, I believe that education and accurate knowledge are the answer, not big-gun government regulation. I think that if some of those products need more adequate warnings on the labels, I don't necessarily think that they should be taken from the market so people don't have access to them.[1]

I personally think that we need to get back to letting people be adults and take some self-responsibility. I think we have tried way too hard to protect everyone from their own foolishness in our society. I have experimented with myself on certain things and didn't particularly like the results, but it didn't kill me. I think paying attention to your own body is a good part of healing.

WORK AS A CLINICAL HERBALIST

One of my biggest problems as an herbalist has been having people take responsibility for their healing. I find that when people come and consult with me, they want me to do what the doctor does. Take charge. Tell them what to do. And be responsible for their health. I won't do that. My job is to teach them how to take care of themselves. I don't use herbal medicine in a vacuum either. For me, I am very much a holistic practitioner. I try to teach [people] to improve their diet, to deal with the stress in their [lives], to do different things. I really believe in the patients' responsibility for themselves. If someone doesn't do that, I don't want [him or her] for a client. I tell the [person] to go back to the medical doctors.

I have been frustrated with some of my clients because I try to explain to them certain holistic ideas, and they can't grasp it. It is a problem. We are creating a paradigm shift in our culture and that never happens overnight. There are people who get it and there are people who don't. I know that right now with this huge popularity of herbs.... I remember how frustrated I was when *20/20* aired the thing on St. John's wort.[2] I had dozens of people come into the shop looking for St. John's wort. I would ask them what they were looking for and why. They would say, "Oh I just heard it would make me feel good." To some of them I would say, "Well, there are other things that could help," because I had worked with depression a lot. So, I said if this doesn't work, come back. It really frustrated me that it appeared that most people have more faith in a ten-minute segment on television than they do in an herbalist with over twenty years of experience. It is a weird thing about our culture. I don't know. I don't know how to solve that problem. I think that is why with the AHG what we are trying to do is to put together educational guidelines for the training of herbalists. We are now working on a national certification test, and a mentorship program so that people can get clinical experience. Our goal is to have recognition for a class of well-trained, professional herbalists, but we don't want to do away with the right of ordinary people to use herbs. It is going to take some public education.

LICENSURE

The AHG has always stood for competence not licensure.[3] Licensure doesn't guarantee competency. What we are interested in is trying to create standards of competency. What we are hoping for is probably more of a certification rather than a licensure, which doesn't exclude other people. It is like this: You have bookkeepers and you have certified public accountants, right? Anybody can be a bookkeeper. You don't have to have any kind of a certification to be a bookkeeper, but a certified public accountant means that you have meet certain criteria. That is the kind of system that we are trying to put in place for herbalists with the AHG. There are some members of our community who want licensure. There are a lot more that are scared to death of that because when you start talking licensure, then you start talking about excluding people, making it a crime to practice without that license. A lot of our members are like me, we don't want to see that person not be able to treat their own headache or their own menstrual cramps, or whatever, with herbs.

I want to mention something, an interesting thing that one of our members had happen in his home state [of] Minnesota. There was a man who was charged with practicing medicine without a license because he was doing some sort of alternative therapy. It went to the courts, and I think it created a real stir in that state so the state appointed a commission to investigate what was going on with alternative therapies. They did this for two years. They basically came back and said, "We don't see any evidence that any harm is being done by these alternative therapies. And where there is no evidence of harm, there is no reason for regulation." They have been drafting a bill of informed consent that basically states that as long as you are not practicing certain things like surgery and prescription drugs and that sort of thing, and you disclose to your client your therapy, and what it entails, that you would be free to provide what is called complementary health care.[4] The deal about complementary health care is that it doesn't involve the right to diagnose diseases, but there should be no reason you shouldn't be able to diagnose under traditional systems that are internal to your own thing. For example, as an herbalist, I don't try to tell [people] that they have cancer or high blood pressure or whatever. They can come to me with the doctor's assessment. I look at things like the tongue, the pulse, the color of their skin, and the description of their symptoms. The model is different. I am not labeling the disease the way the medical doctor does. I think that would be the model. I, personally—and I am not speaking for the AHG—would be pleased with it. I would like to see informed consent with nontoxic, noninvasive therapies.

I see nothing wrong with a person who has been to even a weekend course recommending some herbs for colds, headaches, menstrual cramps, that sort of thing. There are a lot of very common illnesses that you can play around with herbs a lot and it is extremely safe. They are self-limiting conditions. When you get into more a serious illness, it requires a different level of training. Even though I have been involved for years with people who do that, who treat serious illness and try to do it even with limited knowledge, I will be frustrated because I have been a person who has always recognized my own limits. When someone comes to me with a situation that I know is beyond [my] level of training or I don't have adequate experience to deal with, I have never tried to deal with it. I may have recommended some adjunct things to whatever they are doing with the doctor to support their body or improve their general health, but I am not trying to treat that particular disease. There are a lot of people out there who do. And quite honestly, the laws are already in place that would give you the right to punish someone who does that.

AHG GUIDELINES

The guidelines that we have set up with the AHG are twelve hundred hours of instruction and four hundred hours of supervised clinical practice to make sixteen hundred hours. We have included in that some basic sciences, anatomy and physiology, four hundred hours of materia medica and natural therapeutics, four hundred hours of diagnostic skills. The current guidelines require four years of training and/or experience. I think that the twelve hundred hours are sufficient to train a person to work confidently with herbs. We have a lot of schools that are just waiting to set up their programs to match our guidelines. I think six months to a year from now there will be twenty or thirty different schools that will be meeting our criteria.[5] We are getting ready to publish our education directory and there are a lot of programs out there, including lengthy apprenticeships, two-year type things. I think that there is an ability here, if we can just put something in motion, to really have a career path for people who want to become herbalists and start training a bunch of very competent herbalists in this country. That would be a wonderful thing. David Winston made a comment once. He said, there are millions of people in this country suffering unnecessarily from a lack of good herbal care. I agree. There are a lot of things that people are suffering from that could be easily eliminated if they had the proper herbal care.

SCIENCE AND FOLKLORE IN HERBALISM

We are concerned that the paradigm in which herbalism operates will be preempted by the medical paradigm unless we do something to put our own paradigm and our own philosophy forward and add enough Western science so that people feel comfortable and understand that we are competent. The key here is maintaining that holistic paradigm. If medical doctors take over herbs but don't learn herbalism then what we will have is the same thing that happened at the turn of the century. The Eclectic movement was supplanted by modern medicine. Herbalism went out of favor. For example, one of the issues that a lot of us have is that as it moves into a medical model, then what you have is you start with these standardized extracts where you concentrate certain constituents. You say this is more scientific. Well, that is what happened before. That is what ruined the Eclectic movement. In a lot of cases, we know from . . . experience that some of those standardized extracts don't work as well as the whole plant. That is a whole different way of thinking about it.

The history and the folklore are where we actually find the traditional herbalism. There is this idea that we have in our society that our civilization is smart and everybody that ever lived before was stupid. It isn't true. While you do find funny stuff, what you also find is that when people have used the plants traditionally, you can usually depend on that information being fairly accurate. Let's take yarrow, for example. Yarrow was used to treat battle wounds on three different continents for over two thousand years. To dismiss that and say, "Oh, that was just folklore. That was just imagination. All of those soldiers who had yarrow stuffed in their wounds, that was just a placebo . . ." that is awfully arrogant, especially since I have stuffed yarrow in a wound and watched it do exactly what the tradition says it does.

The truth is that scientific thinking tends to be reductionist. If you look at the history of science, every major scientific breakthrough has always been ridiculed and persecuted by the entrenched orthodoxy. There is a tendency for people to want to believe that they know the truth and they have got the answer.

I am not opposed to science when you think of science, as I read in one book, as "accurate knowledge." I look at something as being scientific when I can get duplicatable results. I am very interested in duplicatable results, not random results, results that have been duplicated over many people for a long period of time. That is actually what studying the tradition tells you. It tells you that these people have used this plant for these purposes; multiple people have used it over multiple generations, and reported the exact same result for that plant on the human physiology. This is done by observation of clients, living clients. That becomes a good basis for experimenting today. There shouldn't be any reason why we can't do clinical trials like they are being done in China. If we know that about one-third of all people will respond with placebo, and you administer an herbal medicine in a certain way, and you get seventy percent of the people getting results, you are beyond placebo.

I think part of my frustration with the medical system is that there is a difference in treating the human body as a machine and treating a human being. I think that change needs to take place in medicine across the board, including in people who prescribe herbs allopathically. One of the clinical herbalists that I know who is the best is Matthew Wood. He can tell you how these different herbal remedies affect personalities. And herbs do affect personalities. Herbs can heal emotional issues as well as physical ones. That is one of the remarkable things about the medicine. It is a whole different paradigm. Maybe it is regarded as unscientific because it is out of vogue with the current thinking.

HOLISTIC HERBALISM

The whole idea of vital force, the whole idea that the plant or the whole living thing has a vital force—I don't get hung up on words so I don't care whether we call it chi or spirit. Native Americans would have called it spirit, Chinese would call it chi; but maybe they are not talking about the same thing. But the concept of some abstract thing called life exists and that life is more than a random bunch of chemical processes existing in the physical world. My partner and I are writing a book on pain relief without medication, and one of the things that we were talking about is emotional pain relief. The Cartesian philosophy of separating the body and the spirit, of separating us into pieces, has basically created a schizophrenic culture. You cannot separate the whole. It ceases to exist. You can't look at the heart apart from the liver. You can't look at the disease of the person apart from who they are and the relationships they are in. One of the most interesting questions I have learned to ask as a clinician is—I will get someone who has had a recurring condition for three or four years; they have been battling with it—What happened in the few months prior to the first time you had this? You know how often they tell me exactly what happened? I had a lady who came in and she had recurring respiratory problems. I said, "Okay, when did this first occur? Did anything traumatic happen in your life when that first occurred?" You know what I found out? Her husband had died. And every year on the anniversary of his death, these respiratory problems would come back. In Chinese medicine, they connect grief with the lungs. I have seen this multiple times where an unresolved issue with grief is causing the respiratory problem. I have had a number of cases. I see these connections all the time. To me, they are clear as can be.

You see, we have to stop thinking about human beings [as medical conditions] like, "Oh, you have an ulcer. We will give you this medication." If you take some time to talk, you often find what is causing it. The stress at work, the bad marriage, the poor diet. Then you can treat it holistically. This is one of the things that I think is missing from our modern methods because we want to reduce everything down to a little tiny scientific study, and have everything spelled out so we can . . . everything is not that mechanical. The body is not like a car. A human being is not an automobile that you just replace parts. Yes, I think that is very much more integrated into herbalism than it is into Western medical science. I think it is integrated into it just by the very nature of the plants. If you go out and you identify the plants and you see them growing in nature, you see how they occupy their niche in nature.

For me, it is this Native American concept of medicine power. You see that plant struggle for survival and the challenges it has in its environment. You see how that impacts your body to help you to cope with your environment. It is like milk thistle, which is so popular for the liver. Have you ever seen thistle? What are they like? Don't mess around with me. Traditional medicine says anger resides in the liver. What does milk thistle do? Have you ever tried to kill a thistle? If you spray them with pesticides the top tends to wilt, and they grow back. Does it surprise us that this plant protects the liver against chemical poisoning? Do you understand what you are saying with this holistic way of thinking? Sure, I appreciate the science that shows that this plant contains the flavonoid complex that helps stabilize liver cells so they are less affected by chemical poisoning, but there is something deeper going on. Why does it create this flavonoid? What is there in the nature of this plant that it does that? That is a way of thinking that is whole-brained, not half-brained.

PREDICTABILITY OF HERBS

The longer I have been in this and the more I understand my remedies, there are so many things that I find I can depend on. I can depend on it all the time. It always works. It always does exactly what I know it will do. In other words, the herb always has the same physiological effect. Cascara sagrada always moves the bowels. Cayenne always makes your face flush and gets the circulation going. These remedies have very clear observable physiological effects and what is often attributed to herbal remedies as side effects is actually just part of their effect. It does that. You have to learn what this remedy does to the human body and then you can see when that is going to be needed. You can go ahead and administer that remedy accordingly. For instance, I was talking to a guy who was very scientifically oriented, but he was also open to herbs. I was telling him that there are certain indications; I use this Chinese formula for what I call adrenal burnout, which the Chinese would call a deficiency of fire. There are certain indications that I look for, like the person is tired, but they wake up frequently in the night. They have dark circles under their eyes, their pupils are pulsing, and their tongue is quivering. I know that their adrenals are shot. I see a lot of clients that have this particular profile. I know that if I can get them to lay off the caffeine and the sugar and take a couple of herbal formulas, I have better than ninety-five percent success rates within a few days. I would love to do some kind of controlled clinical thing and document that that actually worked. But what I would have to do is instead of just isolating and treating someone who is tired, I would have to take the client and check [him or her] for how many of these traditional indications [he or she has]. I would put that information into the computer. You could document what happened with the [client]. There are ways of statistically working with that data to find out if there is a statistical significance to those results. But that takes money. I would love if someone would set up an office to fund herbalism with a couple of billion dollars or even a few million dollars. Anything.

Chapter 18

Karta Purkh Singh (K. P.) Khalsa

Karta Purkh Singh (K. P.) Khalsa is one of a handful of clinicians with more than twenty years of experience with medicinal herbs. He is the author of several books on natural healing topics and recently co-authored *Herbal Defense* (1997), a book for the popular market on the use of herbs for staying healthy in modern times. He is a frequent contributor to mainstream and professional publications in the natural healing field, and is a contributing editor to *Let's Live, The Herb Quarterly,* and *Great Life.*

K. P. is a health educator who teaches at all levels of education, from general public presentation to professional training programs. Over the years, he has taught extensively at many community colleges in Washington, Oregon, and Colorado. He is on the faculty of several professional training programs. He is the founder of the Professional Herbalist Certification Course, a two-year, postsecondary curriculum that trains professional herbalists, now being offered at colleges in Washington (Seattle, Olympia, Tacoma) and in New Mexico (Albuquerque).

I spoke with K. P. over the phone on August 19, 1999.

JOURNEY TO HERBALISM

I was diagnosed at age ten with a degenerative spinal disease, which is typically fatal. They told me . . . that there was no medical treatment for my condition and that typically people had gradual degeneration. Most people didn't make it past age forty because as the spine degenerated, the pressure on the nerves was so great that they began to snip the spinal nerves one by one to alleviate the pain. Most people would end up in bed, not improving. At age ten, I got that grim news, but in a way I was fortunate to get it at such a young age. It didn't really register very well. I was also lucky in that there was nothing that conventional medicine could do, so as I became a young adult, I realized that anything that I could do couldn't be any worse than what conventional medicine offered. I decided to investigate some options. I began to experiment with a few things here and there. They were going to do a spinal fusion operation when I was a young adult, which I was able to delay by using some holistic lifestyle practices, yoga especially.

As I began to get more and more interested in what natural healing could do for me, I became fascinated by the whole concept. This was the early 1970s and a time when there wasn't a lot of interest in such things. It was barely beginning to become available. I began experimenting [by] myself and with others. At that same time, I met my mentor, Yogi Bhajan, who told me that I would be a good herbalist. I was nineteen years old and I said, "Oh really?" He said, "If you would like to study with me, I am offering it." I said, "Yeah, that would be great." I have been studying with him [ever since]. Almost thirty years now. I began to experiment with the things he was teaching me and the stuff I was learning by myself. He had me working with people. I began to practice herbalism under his mentorship. I began to see the great things that natural healing could do. I gravitated more and more to herbal medicine as I saw what it was able to do for people. I have been so lucky to see so many miracle stories over the decades. It was really just from personal experience and seeing what could be accomplished that I got more and more interested, and more and more fascinated.

There weren't many practitioners of herbal medicine back then. I was seeing the people who were available, which was pretty limited at that point. I would see chiropractors from time to time. Anybody who was coming through town, I would take their class or talk to them. You have got to imagine what it was like then. In the health food store, you had wheat germ on one shelf and alfalfa tablets on the other and that was about it. It just wasn't well developed like it is now. Acupuncturists? Forget it. It was an age of tremendous experimentation.

HERBS VERSUS HERBALISM

Herbalism is a philosophy or an ethic or a way of conducting yourself in healing that has to do with how these herbs are applied in a healing situation. It is a systematic approach to the use of herbal medicine that is honored in every culture, everywhere around the world.

Every established culture has an herbal medicine that is many centuries old. The herbs are the medicine or the tool; the herbalism is the results you are expecting. In our culture, we are a bunch of immigrants who came here and left the established paradigm of herbal medicine from our particular ethnic background . . . wherever it was. The traditions really didn't get regenerated here very well, certainly in the last one hundred years. There have been various political and social changes. We cast our lot with what has now become modern medicine about one hundred years ago. Herbalism was very dynamic before that stage. How to use herbs in a medical context, in a structured and systematic way—that you could learn academically, and you could learn clinically. It really made sense.

Now what we have in the new herbal gold rush is just the chaos of manufacturers dominating the public consciousness. They are making the products, and telling people they will work. People don't have a historical context for that so they go out and use them and then substantiate that they work. Most people's experience of herbal medicine is pretty skimpy. They get a cold; they use some echinacea. Then they say things like, "Well, it worked great because instead of my cold lasting for ten days, it lasted for eight days." For an herbalist, that is just stupid. We want to see their cold gone in twenty-four hours or we know we didn't do the right thing. You don't just blast away at it with small doses of echinacea. We have to have a holistic context to put this in. What medicine do you use? How do you match it with the patient? What dose do you use? What results do you expect? That is herbalism.

CONSEQUENCES OF A LACK OF HERBALISM IN AMERICA

Since I was trained in that, it is very obvious to me that what we have now is a bunch of laypeople with medical conditions trying to treat themselves using medicine and doing things they would never dream of with drugs, but expecting druglike results. The whole thing is just silly, but people don't know that an organized, systematic way of approaching this exists in our culture. They are starting to now. People have some kind of vague notion that there is something to Chinese medicine and there is something beyond what we know. It is starting to dawn on them. At the moment, it is a mess.

It is not only that people are disorganized, but also the whole theoretical and social framework in which herbalism exists in every other culture isn't here. So, the market is flooded with products that nobody has the experience or sophistication to figure out how to use. The law won't let the manufacturers actually say how to use them properly. They have to use a bunch of innuendo. So, people read a book like mine, and they try to find some product that will match that. They have no way of knowing whether the product actually did what it was capable of doing. They don't have the lifetime of experience with seeing their grandparents, parents, aunts, uncles, everybody else in the village use it that way and get good results.

HERBALISM IN EUROPE

What Europe has, pretty much in all the European countries, is a three-tier system. On the bottom rung, you have folk use. It is something everybody knows about, and you can get those products at the grocery store. The next level is sort of clan herbalism where one member of a group is designated to learn folk herbalism in a much more in-depth and intensive way. They treat self-limiting conditions that can be treated with a small repertoire of herbs that are culturally well known and relatively available, but usually you have to go to some special place to get them. They have some special product designation.

At the top level, you have medical doctors who practice botanical medicine. The medicines are controlled to whatever extent, and they are administered by highly skilled practitioners who treat serious diseases. They are integrated into the medical profession. It is the most developed at the high end, but there are still those other two levels, just as well elucidated. People generally go and get the remedies themselves, and the only level that is controlled to any great extent is the national level. Every country chooses which herbs to control. Echinacea, for example, is controlled by prescribing practitioners in Europe. For whatever reason, they choose particular remedies that they want to control. That varies from country to country.

AMERICAN SYSTEM VERSUS THE EUROPEAN SYSTEM

In America, the market appeal or the social appeal has been from the grass roots—people like me and my students and my patients, people who read herb books and the baby boomers from the 1960s who now have grown up to the point where they are starting to have serious needs in terms of health and they want to do it in some alternative way. The advantage of a more formalized system is that people are going to get much better clinical results. There will be much less misuse of popular herbs. Here, probably ninety percent of the echinacea that is taken is taken improperly, for the wrong uses, in the wrong dose; the quality is bad; and it is clear that it is an ecological catastrophe with that particular plant. People are scarfing down tons of echinacea with no benefit. It is silly.

The potential disadvantage of this situation is that the grassroots people will lose access to some of the things that they otherwise can use wisely and appropriately. Hopefully, we will find a way that it won't happen that way. People will be able to understand what their limits are. Anybody should be able to brew up a cup of peppermint tea for their tummies. Probably anybody should be able to take echinacea for a cold if [he or she] learn[s] the appropriate use for it. If we could educate people and put on the package how to do it and what it is used for, it will become culturally integrated. Then there are definitely certain things that should be restricted, an herb like belladonna, the deadly nightshade, which is used in doses of a drop of tincture. That is a very valuable professional tool, but I don't want the average person out there having access to it. There are advantages and disadvantages.

I think that the European system is heavily influenced by contemporary medical practice, and they tend to have a "one disease, one herb" mentality. They are creating very druglike herbal substances. A great example is the herb black cohosh that I have been using for my whole career. A simple herb that grows in Appalachia, you dig it out of the Earth. People have been doing it there for seven generations. You dry the root and put it in a capsule. It treats hot flashes very well. Now, for whatever reason, black cohosh grabbed the attention of the European researchers. They concentrated it, and they have standardized the extract, standardized it to a marker compound, not even the active ingredient. They published a bunch of studies. Everybody got real excited about it. Now American herb companies have started to reimport the German high-tech standardized extract at five times the price that it would be if we just took the doggone stuff we dig out of the ground in Appalachia. This herb got a whole lot of attention. We are raping the land like we did with echinacea, digging out every last bit of black cohosh root and shipping it to Europe. They put it through a bunch of high-tech doddling to create something that works no better than the herb did in the first place, but you get to pay more for it. Now, it is the herb du jour because people can read all of these German studies. That is a disadvantage of the way the German system works. I was just talking to someone today who is setting up a deal to import fifty thousand dollars worth of these black cohosh tablets. I said, you want women to get a good buy on this stuff? Instead of buying this German stuff, just buy the plant material. You will still have great results. This is where a grassroots understanding would be valuable and sort of a Wise Woman approach to how can we save our ecosystem but still use herbal medicine that has minimal side effects, that works well for people, that costs a lot less.

BENEFITS OF REGULATION

I think that there are herbs that we do want to be out of the reach of the common person. If we had enough sense not to mess with those herbs, I would say great. I think we are only years away from the first case of belladonna poisoning or aconite poisoning or arnica poisoning because there is not enough common sense out there among the people. Eventually somebody is going to get ahold of something like that. My sense is that government regulation is probably the only way to allow those to be handled safely, as much as I would like it to be a medicine of the people and be unregulated. Historically, it has certainly been that. Although, when you look at cultures that have not been as influenced by contemporary modern medicine and who have well-established systems of herbal medicine, mainly in India and China, you have a really well-established, essentially three-tiered, system that has worked really well. People are really respectful of it, and they know that if they have indigestion, they can go down to the herb store on the corner and brew something up. If they get persistent menstrual cramps, they can go to the grandma who lives down the block and treats everybody in the extended family with medicine that she knows about. But if they get chronic asthma that is not being relieved by grandma, then they can go to the Ayurvedic doctor who uses even more powerful medicines that are out of the range of grandma's education. It just works really well. The system is pretty informal, but it serves people well.

LICENSURE AND PROFESSIONALIZATION

Professionalization is a very hot topic among professional herbalists right now. We want to have appropriate access for people, based on their level of training and experience, to herbal medicine across the board. Not only do we want people who have no training to not be able to practice, but we also want to keep things available for grassroots use. It is a tricky balance. We are really scrambling with how to figure this out. But frankly, every profession has gone through this. When acupuncture first began to formalize, the initial acupuncture board was all medical doctors who knew "bubkes" about acupuncture, but those were the ones that the legislature allowed to establish the board. Often, you would have a Chinese doctor with thirty years of experience being examined for a license by a board of five medical doctors who had taken a weekend seminar. But now, as we have achieved some balance with that, that is no longer the case. Now you have acupuncture boards run by acupuncturists with maybe a token medical doctor.

We have definitely seen a change with herbalism even over the past year. A year ago pharmacists as a whole were motivated to take over the dissemination of herbal medicine. That was definitely their orientation. They were training pharmacists in herbal protocols. Pharmacists were thinking that they would be the logical ones to do it. They don't have prescriptive power in medicine but they could essentially prescribe how to use herbs over the counter. After trying to do that for about a year, they are now realizing that the body of knowledge is incredibly different. The paradigm is so different, the materia medica is so different, that they don't have the tools to be able to do that without going through another complete education, without becoming educated the way I did in addition to the way they did. Some of them are doing that. We just had our professional conference last weekend.

About half the people attending were medical professionals, which is a big thing—medical doctors, nurse practitioners. We had several people who were directors of integrated care facilities of major hospitals, and they were there to figure out how they could interact with professional herbalists. Basically their message was, we thought that we could just learn your medicine, but when we began doing that, we realized that your education had been as in depth as ours. We need to educate ourselves from the bottom up if we are going to be prescribing "herbal medical doctors." Maybe it would be better to strike an agreement so that you're the botanical medicine specialists, and we know enough to know when we should refer. We will give people ginger for their tummy aches, but when we want to treat asthma with herbal medicine, we will have you in our clinic and refer to you. It was a big shift in consciousness and it happened really quickly. Not all medical doctors think that way. I was concerned about orthodox medicine co-opting herbal medicine, but from what has happened even in the last six months, I am less concerned. I think they are being pretty honest as a group about what they will do.

PERSONAL REASONS FOR SUPPORTING LICENSURE

I would love it if we would just, what one of my colleagues called, legalize adulthood—make people responsible for their personal decisions and allow people to contract with any health care provider that they choose as long as there was full disclosure. That is not the way we tend to operate here. The Netherlands, for example, has that sort of a policy with health care. They do well. We are a much more litigious society. We tend to regulate. Our way of handling that is to regulate. Top-down solutions. As a person who has been in clinical practice for twenty-five years, I want to be able to participate in the banquet of health care services that Americans are using. I want people to be able to see me legitimately, aboveboard, to be able to communicate with [their] other practitioners, and have people be able to afford that service which means third-party payment eventually. That is a real sticky point for people. Their insurance is covered by their employer. There is a certain amount of out-of-pocket expense that they will tolerate, but there is a limit. We need third-party payment. Just in the last six months a bunch of major insurance companies have begun covering the botanical medicines themselves. They were very resistant to doing this. They have to be prescribed by a specific category of practitioner. It kind of got back to that problem with the medical doctors prescribing herbs, instead of an herbalist. The way they are getting around it right now is that they bring herbalists on staff as consultants. Herbalists don't do the prescribing. They evaluate the case and then consult with the medical doctor or the nutritionist or whoever is the one who is officially providing care, who has the prescriptive power. Then, they are paid as a consultant. It is working out in an informal way so far.

I think that licensure is inevitable. I have seen it happen in every other holistic professional field. There is a lot of argument about exactly what form it should take. Every state does it differently. I think that the argument that herbal medicine is the medicine of the people, and it shouldn't be restricted is kind of a spurious argument because we could say the same thing about nutritional consulting, massage therapy, fitness trainers, a lot of people who have skills that, on the lower level, anybody could do or learn. I don't think that anybody is thinking that mommy is not going to be able to give her kid peppermint tea for his or her tummy. I think it is a spurious argument. If I give my wife a back rub, I am not being arrested for practicing massage without a license. However, I don't want untrained people out there giving out arnica tincture. I think some regulation is appropriate, and if we could do it through some kind of common understanding then great, but I just don't see that happening. Herbalists need to have hospital privileges. If an herbalist is treating a patient and that patient goes into the hospital, that herbalist is needed to manage the pain. If [herbalists] can't be in the hospital, then there is a resource for the patient which is gone. In order to have hospital privileges, they have to be legitimized. There you go.

POSSIBLE RIFTS WITHIN HERBALISM DUE TO LICENSURE

Herbalism is sort of the last gasp of the dissatisfied, alienated maverick. They have been frozen out of massage therapy, acupuncture, nutritional health and dietetics, chiropractic, naturopathy. We are down to herbalism as the last place that they can practice without fear of reprisal. We somehow in being behind all of the other professions in becoming legitimized have gotten the grab bag of the leftovers. But frankly, all these other professions have gone through exactly the same issue. Massage therapy had the same . . . "you know, it is the God-given right of the people to practice." We don't need regulation. All of a sudden a bunch of people are getting agitated. "Why have Big Brother looking over our shoulder." Yes, I am sympathetic to all of those arguments. I understand them. I just think that pragmatically this is the reality in our culture; this is the place where we draw a line in the sand. It is going to be in our best interest, just purely because that is the way we do things here. In other countries, where things are a little looser maybe that would be justifiable or in other countries that have a lot more ethical and personal responsibility like the Netherlands, there might be another model that would work well. It just always seems funny to me that people somehow feel this argument, that it is a medicine of the people, and that right shouldn't be abridged, somehow are focused on herbalism. Every other aspect of our health care is in our hands and there are lay people participating and there are these tiers of expertise. I don't know why that argument should apply to herbalism and not to dietary consulting. Why do we have registered dietitians and licensed nutritionists? Maybe they made the same argument. I don't know. The food we eat is just as basic as the herbal medicine that we take.

I think that more established mainstream groups will take up the cause and co-opt herbalism if we don't organize. I don't think they want to at this point. The medical doctors did. The pharmacists did. They are now realizing that maybe that wouldn't be such a good idea and wouldn't work very well. But given a vacuum, any of those groups could slip in if herbalists aren't a potent political force. Frankly, I am not too concerned about that though because reactionaries tend to be a split off, splinter group that does its own thing. There is a whole group of unlicensed acupuncturists, who were disenchanted when acupuncture became licensed, who practice unlicensed. I live here in Seattle. It has a very large Asian community and I guarantee there are as many Asian acupuncturists practicing unlicensed as there are licensed. In their case, it might not be any grand philosophical statement. It is just that they don't speak English well enough to pass the test or whatever the deal is. Everybody is not all excited about that and going out and calling that a split in the ranks. You still have licensed acupuncturists who seek insurance reimbursement. They have schools that teach people, and more and more people are becoming licensed. Eventually those malcontents kind of slip away. I think, at the beginning of legitimization, every profession has experienced this. There you are. I think chiropractic has been around long enough that those malcontents no longer exist. I don't see a big underground movement of unlicensed chiropractors, but there certainly was at the time when they became legitimized. They had the same fight. So, yes, I think that there will be some squawking. But so far this is the direction every profession has gone.[1]

HEALTH CONSULTING IN THE NATURAL FOOD STORE

I put on earmuffs when I walk into a health food store. I hear some of the most chilling conversations when I am there.

The issue of therapeutics in the marketplace is a very tricky one. Frankly, I am conflicted about it. If the minimum-wage, health food store clerk is giving people medical advice, I would rather that they give them the most educated medical advice that they can. On the other hand, I don't want them giving any medical advice. It totally frustrates me because I would send my clients to the herb store only to have my advice contradicted by a nineteen-year-old who has been there for a week. I would call up and howl at the owner of the store who would promise it would never happen again and then it happens the next week. Store clerks are in that place because they are fascinated by herbs. They just can't control themselves. It is not legal. That is true. They should not be doing it. It is such a gray area, though. They have as much legal right to do it as I do actually, depending on how they phrase it. If we talk in that kind of legalistic language where we are not quite saying something that we are not supposed to say, then the question of legality is again kind of a spurious point. I am more concerned about their qualifications. Some of my colleagues are very well-trained clinical herbalists and work in herb stores or own them or are herbal consultants, in that way. Those people are giving tremendous herbal advice. Others of them are college students and part-time counter people.

I am a great proponent of information. I think that the more information everybody has about everything, the better. Full disclosure is always the best policy. If they are going to prescribe something at the health food store, better that they have some training, in fact, there is probably a lot that they can learn that they can do very effectively if they knew their limits. We should create some kind of credentialing system for those people. One of the reasons this isn't working so far is the economic incentive. The law allows anybody to be in there and the health food store owners overtly claim that they are not—that their counter people are not giving medical advice, but covertly actually allow it to happen, or even encourage it, because it makes sales. There is an economic incentive for them to hire the cheapest person but also have that person give the most advice and sell the most product. Unless that is going to get controlled legally, I think the economic incentive is too great. That is where consumer education ultimately has to take place. People have to start demanding to know the credentials of the people giving the advice. I teach a two-year training program for professionals, for example. At the end of the first year, that would be a very good level of education for people to go into a health food store and be able to make suggestions about non-life-threatening conditions. That would be the kind of credential people could demand to see. Health food stores are getting more interested in having more credentialed people, although they are still reluctant to pay them.

NEED FOR CONSUMER EDUCATION

We have herbs but lack herbalism. Herbalism is a system which is infused through an entire culture, so the consumers understand their limit and the practitioners understand their limit. You don't have this problem in Europe because people grew up with herbal medicine. They know what it can do and what it cannot do. They know what conditions are effectively treated. They will go in and buy the old family remedy that the grandparents taught them about, but if they get a howling ear infection they will go to their professional herbalist, who will have some more powerful herbs that they can use. It is just part of their culture. Here, people go into the natural food store with a bleeding peptic ulcer and ask some "nudnik" what they should be taking and believe them. It is scary.

The people I would really like to educate are the consumers. I train professionals primarily. That is what I spend most of my time doing these days, but where I would really like to see the educational focus is on the consumer. I would explain to them what is reasonable. What they can treat. What they can't treat. I encourage people to not even interact with the clerk at the health food store. Educate yourself. Figure out what you want to do. Go in there, get it, and get out. Don't let them even engage you. The average person knows more about [his or her] own medical condition than the minimum-wage counter person knows. That is where the education should be. As people read more and take courses . . . it will take a generation until we infuse our culture with this knowledge of what is what. My seven-year-old daughter is already a pretty good herb expert. In other cultures . . . I heard a colleague of mine describe the other day, people are born having this knowledge in their bones. It is really true. You see grandma and grandpa using it and everybody on the block. You really get to know how it is going to work and what you do and what you can't do, and when you see grandma and when you see the professional. It is a whole cultural paradigm that has developed. It is going to be really sketchy for a generation in this country.

The whole system is set up to perpetuate itself, though. The reality is that the health food store is selling medicine in a nonmedical way. Of course, people are having their medicine paid for by insurance and there are only so much natural products that they will buy. If the health food stores pay their employees more, then the price of the products goes up. It goes round and round. It is probably going to gradually change, as some of the forces come into play and you see people exerting more and more pressure to know who is giving them advice. It is shocking to me the kind of advice that people will take. That is one of the problems in our culture. We simultaneously hold these conflicting dualities, essential dualities. One is that herbal medicines are powerful and effective and will treat disease, yet we think they have no side effects. It can't be both. They can be one or the other or the reality is that they are both but not in the way we think. People do things with herbal medicine that they would never do with drugs because they are able to convince themselves that herbs are not drugs. Yet they expect them to perform like drugs for their condition. It will just take a while to get all of this stuff straightened out.

ISOLATION OF HERBALISM AS A HEALING MODALITY

One of the problems with defining herbalism is that herbalism is a modality, a way of choosing how to treat conditions. In our culture, healing specialties tend to be divided up either by practice, definitions, or by modalities. So, in Washington we have a credentialed naturopathic physician. It is very popular here. It is kind of a national center for naturopathy. People are very aware of it here. They accept them as part of the health care system. They are paid for by insurance. They have a very large scope of practice. The only thing that identifies them is their scope of practice. What that means is that they are trained as generalists. Any naturopath is going to have minimal training in botanical medicine, but also twelve other healing modalities. They are jack-of-all-trades, master of none. They have no specialty. Some of the best herbalists I know are naturopathic physicians, but most naturopathic physicians are minimal herbalists.

Herbalists, on the other hand, identify themselves based on a modality because they use plant parts for healing. There are all really artificial definitions, though. In most other cultures, traditional healers do a bit of everything. They function in a culture that offers a bit of everything. An acupuncturist doesn't try to be all things to all people. They treat things that are appropriate for acupuncture. They may also be an herbalist and a dietitian. They also might have psychological counseling as part of their repertoire. They try to match the people to the constellation of modalities that are going to match their needs. Herbalism doesn't exist in any other culture by itself. So, it can't here. If you bash yourself with a hammer, you want to do things other than take herbs. Since the development of conventional medicine, natural healing can't exist as the only system of healing. We need trauma care. That is the main one. But I got to say, as an herbalist, I have seen such tremendous results with herbal medicine that there is darn little that it doesn't do a really great job with. But you have to have real specialized training in using the herbs in various ways. Herbal medicine is the equivalent of modern antibiotics in terms of effectiveness, but herbal medicine needs to be used aggressively and with great care. Not everyone who takes antibiotics can cure his or her infection, and not everyone who takes herbal medicine could cure his or her infection. If you choose to use antibiotics for your infection and you die, everyone feels like at least you used the most powerful medicine available. But it didn't work. If you choose to use herbal medicine, you're out of the mainstream and it creates doubt and unsurity. The herb may not have been able to cure your infection.

I had a friend who was a martial artist. In one of his workouts, he broke his forearm. He treated his broken arm with herbal medicine internally and externally and manipulated the bone on a daily basis. He had no cast. He had a very light splint and a sling. Three weeks later he was back in full martial arts workout. He had his arm x-rayed later, and they said his arm had healed as effectively as any arm they had ever seen. He had no atrophy, no loss of function, and it was completely healed in three weeks. Nobody here knows how to do that kind of manipulation anymore because we just send people to go get a cast. Even though it is possible, somebody could be trained in it . . . What is the point? We don't experience the full spectrum of the potential of natural healing here because we automatically use conventional medicine. On the other hand, I broke my toe a while ago. I went straight into the emergency room. My little toe was sticking out at a right angle to the rest of my foot. I had it anesthetized and set. I couldn't imagine doing that without the anesthetic. In the old days, a person would have gotten good and drunk, or [he or she] would have taken opium, or whatever, and it would probably have worked just as well, but I was really glad it could be done conventionally with local anesthetic. That is a great development. I went home and treated it further with herbal medicine and that was that. It healed up fine. So, herbal medicine could exist on its own. Should it? No.

CURRENT AHG VISION OF PROFESSIONALIZING

I expect to see a tiered system develop. Nursing is a great example, where you have a nurses aide, then you have a practical nurse, then you have a licensed practical nurse, then you have an RN, then you have a nurse practitioner. People with different educational levels are given different responsibility. The nurse practitioners have almost the same responsibility as medical doctors. It is a good model to look at. Can a person learn some really useful tools to help . . . maintain his or her health in six months? Absolutely. But they need to know their limits. Someone trained in massage therapy for a year doesn't become a physical therapist that has four years of training. The standard that we have set now . . . the standard that the American Herbalists Guild has established for professionals, for someone to have a practice in herbal medicine as a general practitioner, the equivalent of a primary care provider, is four years of training and/or experience. A combination of those. I expect to see that more formalized over the next generation. Right now, we don't have much hope of being legitimized as primary care practitioners because none of the schools offer the differential diagnostic training. Someone has a headache; you better make sure [he or she doesn't] have a brain hemorrhage. You have to be able to know that difference. We don't have the opportunity for our students to learn that yet, but as a complementary provider . . . we could be in the same category as dietitians, acupuncturists, massage therapists. Our professional standard is four years, which is the equivalent of what a primary care practitioner would achieve, but our education is not formalized enough yet. It is loose.

The way that our association does it now is by peer review. A panel of five established professional herbalists review the applications, and the basic criteria are four years training and/or experience. The reason for that is herbalists immerse themselves in herbs. It is the equivalent of four-year full-time work. If you went to herb school in England for four years and you have your diploma . . . that is the kind of standard that we are looking for. There are no American schools like that. People are trained by apprenticeship and mentorship and self-study and all kinds of combinations of those. We are looking for that level of expertise.

I teach a program which is two years, essentially two years of full time. It is all academic. Then people have to go out and do clinical training. We have a student clinic that offers 100 program hours which is about one-quarter of what we like to see people have. They should have four hundred clinical hours. Students are supervised by professional herbalists.

We are just at the beginning of creating standards of practice. They will be formalized as time goes on. We are in the process of creating national certification standards so that people can be admitted that way. It is tightening up. As we become more and more a part of the scene and more and more legitimized, you are going to see those standards change.

OVERALL VIEW OF AMERICAN HERBALISM

One thing is that herbalism really works. In our culture we have an idea that herbal medicine is a convenient fantasy. Wouldn't it be nice if it really worked? But everybody knows that with a serious condition, you have to resort to medical practice. What people don't have is the experience of herbal medicine really working for serious clinical conditions. I have spent a quarter of a century working with people who had hay fever or food allergies that they were so debilitated they couldn't live their lives. Now for fifteen years these people haven't had a problem.

Herbalism is really very powerful. It works. It is dynamic. It is dramatic. It is effective. It is dealable. And it is not just, take a couple of echinacea capsules. People need to have that contact with that serious side of herbal medicine. That only comes when we reach a critical mass of having enough people willing to experiment a bit and meeting other people who have had that kind of result. I, myself, am able to stand here today because of the power of herbal medicine. I have seen these kind of results so many hundreds of times . . . it really works.

Chapter 19

Corinne Martin

Corinne Martin is a clinical herbalist and teacher based in Portland, Maine, and has been working with plants for more than twenty years. She feels particularly drawn to the use of plants in relation to ecological concerns (i.e., ways to connect people to the land), social justice issues, and a sacred practice of creation spirituality. Corinne has written one book and numerous articles on the topics of herbs and health care. She has also worked with homeless people in a medicinal herb garden project for many years. Corinne holds a master of arts in ethnobotany and sustainable communities as well as a master of divinity degree. She is currently an instructor of holistic health at the University of Southern Maine's College of Nursing and Health Professions.

I spoke with Corinne over the phone on September 13, 1999.

DEFINITION OF AN HERBALIST

I think part of the difficulty with the term herbalist is that it is a very broad definition. That is to say, I know people who would call themselves herbalists, and what they do is grow herbs. I know people who call themselves herbalists and who make products. Other people consider themselves herbalists and are clinical practitioners. I think that when I first was practicing, for me, an herbalist meant a clinical practitioner, someone who worked directly with people on health care issues. Now I meet more and more people who are involved with herbs in different ways. In a sense the definition of an herbalist is becoming more encompassing. I guess if I have to say, I think an herbalist is someone who is involved in working with plants, probably having something to do with healing plants, but not necessarily.

I am thinking of a friend. She has been gardening with herbs forever but does not clinically practice. She uses herbs a little bit for herself and makes her own tinctures and preparations and whatever, but she is basically a gardener. I certainly would not exclude her from the title "herbalist." My definition has broadened over the years. At the same time, I also think we are being asked to refine it now. With the increasing popularity of herbalism especially in this country there are health care practitioners, conventional practitioners, or what people sometimes call allopathic practitioners, who are hoping that herbalism can somehow present itself in a much more defined manner.

INTERFACE OF HERBALISM WITH MEDICAL SCIENCE

Some of my experience working with conventional health care practitioners is that they are really interested in herbs, but they really don't know how to include them because the paradigms are very different. What I am comfortable with, as an herbalist, is something that they are not comfortable with and visa versa. We are trying to bridge paradigms, you know.

In order to build these bridges, we need to continue sharing information. I think that a number of conventional health practitioners are surprised when I start talking, and they find out that I am well informed, not only about herbs but also about health care. I think that is part of it. We have to share ourselves, get out there and talk to people and be willing to exchange information and let them know what premise we are operating from.

We need to share research. As much as most of us, or many of us who are herbalists, want to be out in the garden or out in the wild harvesting or working with people—I think we must at least point conventional practitioners in the direction of research that will be helpful for them.

I guess, though, in part, the links, the bridges get made on their own just because people are going to physicians and saying I am taking such and such. I have people who come to me as clients, and they say, "I told my doctor I was not going to take such and such anymore so I want herbs for it. I told him I was going to find an herbalist." In some ways, those links get formed on their own.

SCIENCE AND THE TRADITION OF INTIMACY IN HERBALISM

It doesn't make me feel totally secure when I can pull up scientific research on the computer. I am much more secure when I can also talk to people who have used the plants and find out what their experiences have been.

When I talk to a health care practitioner, the whole issue of how we evaluate a plant, when a plant is so different from an isolated compound in a plant or an isolated compound that comes out of a laboratory, comes up. Herbalists need to let them know that we are dealing with apples and oranges and that the evaluatory methods for one paradigm don't really suit the other. We just have to keep asking that question, How do we do this? How do we do this integration? That is something that really hasn't occurred to conventional health care practitioners. They assume that the scientific evaluatory method is appropriate for both substances, and it isn't. Just the issue of compounds that can work synergistically is often beyond medical science. For example, you can take one particular plant and it can have carcinogens in it, but it can also have antitumor properties. How do we make sense out of that in light of a pharmacological study, or in light of trying to evaluate an herb with pharmaceutical drug criteria? Those are questions that medical science is not even aware exist. Herbalists need to continue bringing those issues to light.

Another thing is that, as herbalists, we really have to sort of give up the idea that our way is right. That doesn't necessarily mean that we have to pretend that we don't believe in herbs or that we haven't experienced success with herbalism, but for there to be a real bridging of paradigms and a real communal effort at whole healing, neither party can stand on the soapbox or at the podium. We both have to sit at the table and say: We are both working toward whole health. How do we do this? What is your perspective? And what is my perspective? We must be really willing to reevaluate our own perspectives as well. None of us has the whole picture. If we are really going to work together, then I think we really need to let go of our righteousness or our claim to truth and try to find it together.

Generally, I think what ends up happening is that the person who is outside the circle or the party tries to speak the language of the party inside the circle so that a kind of communication can happen. Of course what always ends up happening is that there is something that gets sacrificed. Maybe that is okay and maybe it's not. I think that some of the focus on science and pharmaceutical method and evaluation is at least in part driven by an attempt to have herbs seen as a valuable, effective healing modality. In that sense, it is positive. I also think science has a lot to offer to us as herbalists. It is amazing to figure out why aspirin works or why willow bark works. That adds information for us.

At the same time, on the negative side of it, the paradigm out of which herbalism has worked for so long emphasizes the responsibility of the patient or the client and also offers a lot of space for individual variability. None of this can be scientifically accounted for, nor is it present in the allopathic model of practice.

The sense of intimacy involved with herbalism is a piece that stands a big chance of getting lost in the building of bridges to the biomedical community. Intimacy with the land. Intimacy with the client. If an herbalist practices, then the herbalist is generally growing or harvesting plants, so they are intimate with the land. They sit with a client for an hour, an hour and a half, and [he or she is] intimate with that person. It is a whole cycle where every step of the circle of healing is intimate and tangible. That aspect of it is something that is really crucial and critical in terms of our sustainability as healers and the Earth's sustainability.

I think that as we gain . . . as more distance . . . as there are more pieces and more distance between the pieces, the potential is there for something to get lost that is really critical to herbalism, and I think to healing in general. I hear practitioners talk about it all the time. A doctor can see sixty people in one day, whereas I don't know any herbalist who can see that many people in one day. I know that will change to some degree and yet I am hoping that we can somehow maintain a kind of vision of the necessity for that kind of capacity for intimacy with the land, intimacy with the patient or client, intimacy with other practitioners. When the connections are that tangible and direct what we end up with is a lot more whole.

I am doing a lot of reading in ethnobotany right now. It is very interesting. Some of the research that is being done has to do with looking at land-use management strategies and how much more sustainable these strategies are when the connections are very intimate. It speaks to how ill-equipped the "scientific" means of, say, producing food are in terms of sustainability. When those intimate connections are maintained, sustainability seems to be, at least from some of the things that I have read, much more available, much easier to achieve. We are looking at the sustainability of peoples and sustainability of health, as well as of land, when we look at herbalism.

EDUCATION FOR HERBALISTS

I think it would be terrific if we had two- or three-year programs in accredited schools or schools of some sort where people could get a real background in anatomy and physiology and some biochemistry and nutrition, as well as herbal studies, field studies. I also think that looking at issues like working with people and the place of spirituality in healing are important in there.

When I suggest a study of anatomy and physiology and biochemistry, I am talking about Western herbalism in our country as it stands now. There are of course traditional healers in other countries who have never studied anatomy or physiology in their whole life and practice successfully. If I were designing a curriculum, those things would be in there partly because I think the body is fascinating, and partly because if we are giving herbs, then we need to understand how the chemicals interact in the body.

INTEGRATING IDEAS OF INTIMACY AND SPIRITUALITY INTO HERBAL EDUCATION

I remember attending an herb conference a few years ago. I was listening to a naturopathic practitioner talk about a particular herb. I asked what part of the plant is used. She said, "I don't know. It comes in a little bottle." And I thought, *Huh, isn't that interesting. What an odd way of using herbs.* It just never occurred to me that people were using herbs that had never seen the plant. I think that getting people outdoors and actually doing the practice of identification and harvesting and preparation is part of it. And also looking at other cultures where people live intimately dependent upon the land for healing plants, like native peoples, is really important.

I think the issues are difficult because we are in a difficult time. We are in a challenging time. For herbalists, I think it is very exciting in lots of ways, and yet it also raises a lot of questions.

PROVIDING ALTERNATIVE HEALTH CARE TO THE MAINSTREAM

I think that accessibility for the mainstream is really important. I wouldn't withhold an herb from [people] just because they didn't want to live the way I want to live. That doesn't make any sense. If people can start out thinking about plants because they have used it in a capsule or their physician suggested it and it worked for them, then yippee. The concern about physicians giving the herbs is that they may not be necessarily trained. Maybe they don't come out of the same paradigm or worldview or whatever that I might. If people are interested and they can use something that is less toxic, why not? Somewhere along the line, they are going to figure out that it is about a plant.

In order to get knowledgeable herbal health care to the mainstream, though, herbalists need to let themselves be known. Do some sort of education, invite yourself to speak at a medical practice, either an office or a hospital, and try to give people information about what herbalism is, or sponsor a conference of alternative practitioners. Where I live, there are a number of hospitals that have all kinds of alternative days or fairs. That is one way to help people gain exposure.

Another thing is to become very familiar with all of the resources that are out there and maybe put together a list of resources on herbal medicine or healing; a bibliography or whatever, and get that to hospitals.

The other thing is to be very integritous in terms of your own practice. If people know that we are not crazy and know that we are fairly well informed, then that speaks pretty well for making the link by itself.

I remember speaking to an herbalist once who said that the hospital called her and said this person had such and such and was in the emergency room, and she refused drugs. What herbs would they suggest? The woman didn't know the words that they were saying. And you know there are a million medical words that you cannot know, but to at least try and familiarize yourself with that language I think is really important.

REGULATION OF HERBS

The FDA has the power to say this is the only way you can use these plants, only if such and such research has been done and then only physicians can prescribe it. We have a system in place that generally likes to eliminate any competition. I can't imagine really though that they would run the whole show. The bottom line is people are going to be growing herbs and reading herb books. Herbalists are going to be out there somewhere. I think that herbalists will continue to exist. We are just going to have different access to people. I think we will continue to be there. Enough people are interested now. People are beginning to make connections between nature and healing—healing their own bodies and healing the planet. Because of that, I have a feeling, it's just a personal feeling I guess, but I don't think it will be subsumed any longer.

LICENSURE FOR HERBALISTS

I think for many people accountability is a major factor. Licensure at least makes people comfortable using medicinal plants that might not be comfortable otherwise. The difficulties, of course, are who makes the determination of who gets licensed, and what do we do about people who don't fit the structure that gets set up. I am assuming that if we wrestle with it long enough and with enough integrity, we will come up with something that is viable for all participants.

I think those who will not be licensed will continue to practice as they always have. I remember talking to Deb Soule at one point, and she was saying that the herb community basically regulates itself. The word gets around about who is a good herbalist. That is the community aspect of herbalism that has always been there. You hear things about people. You hear things about certain herbal teachers. The word will get around in terms of who is good and who is not. When whatever licensure system gets created, there has to be categories or slots or accreditation assigned to people who are folk healers, curanderos, or native people who have been practicing and are not going to go to school. Somebody who has been practicing midwifery for thirty years on their own isn't a person without information even if they never went to school. I am assuming that because herbalism is fairly eclectic that it will create categories for people like that. I don't know how the U.S. government will handle that, but I guess we will just have to find out.

COMMUNITY GARDEN PROJECT

Five years ago, a woman contacted me who was a psychiatric nurse practitioner. She was working with a homeless health care clinic in Portland, Maine. She had attended a workshop that I did where we made herbal preparations. She had this dream of having a clinic full of herbs. She said that she had started talking to some of the clients at the clinic. Many of them were what you call dual-diagnosis patients. They were in recovery from substance abuse plus they had psychiatric disorders. Because they were in substance abuse recovery, many of them didn't want to use drugs for their psychiatric problems. They were afraid of becoming addicted. She wondered if herbs might be a viable possibility for their health and psychiatric needs. I said sure, why not.

We attempted to have a little herb clinic where I would come in once a week or once every other week. That was difficult for many of the homeless people because they are psychiatrically disabled . . . and were not able to show up very much for appointments.

Then one of the clients had this idea that we should grow the herbs. Supply was an ongoing issue for us. How do we get these herbs? I would buy them one week, and the clinic director would buy them the next week. Or I would provide whatever I had, but supply was an issue. One of the clients had the idea that we should grow a garden. It turned out that right next to the sleeping shelter there was a lot where the city used to dump all of the trash that they cleaned from the streets. The city volunteered to give us this lot to use for our garden. We got a core of volunteers, partly people who worked in the homeless health care clinic and a couple of herbalists and some homeless residents of the shelter. We got together and cleared up the lot and planted the first garden. We wanted to grow medicinal herbs, but the clients also said that they would like to grow vegetables as well. What we were trying to do was not only grow medicinal plants to use, but also to create a space in which clients who had no place to call home could have a place to do earthwork that would be healing for them.

Every year it got a little bit bigger. We are up to three gardens right now. It has never been easy. It is always difficult because the population of clients fluctuates. They may be here today, gone tomorrow. The neighborhood is pretty terrible. But lots of amazing things have happened. The garden always looks beautiful, every year, no matter what. They always get lots of food. What has happened over time is that there is so much produce that they give it to the food kitchen and the food pantry.

We are now at the stage of trying to form a cottage industry where we are going to be making some salves to sell at the public market. I now have this idea of pulling together homeless people and low-income people. We would train them and then use them as a workforce for which they would get a stipend. We would work on the farms of organic farmers who are hoping to use herbs as an agribusiness or a supplementary crop. Then, we have a number of different players who are getting their needs met and forming greater community in a sustainable way.

I think this year we ended up having three or four core homeless volunteers and others who come if we have a big workday. We have gotten to the point where we pay them a stipend. Either the social services department pays them a stipend to go through our training program and continue to participate, or we give them a ten-dollar gift certificate to the local supermarket for a certain period of work.

The first year that we set up the garden we got somebody to build a cold frame for us. Well, it lasted twenty-four hours. Somebody ripped it off. Sometimes we will find used condoms draped over the plants. Then we had an active schizophrenic man who thought that he was pruning the tomatoes and he went and chopped the tops off all of them. We put up poles for the beans, and he thought that they were pagan symbols so he went and rearranged them all into Christian crosses [laughs]. We have people who are very unstable in their lives and so we might have a crew of eight people who say they are going to show up, and two people actually show up. Those sorts of things.

One of the most difficult things is that we are in Maine and so our growing season is four months out of the year. So, we form intimate connections with these people, and they sit in meetings and talk about all kinds of things, and they have an equal voice in what happens. Then at the end of the growing season, they are back out on the streets. There are a number of things that we haven't figured out ways to address yet. We are constantly struggling with all of those things.

For me, it has been a real experience in terms of trying to teach. Normally I teach people who want to be herbalists or they want to really find out about herbs, so they come and expect to sit down in a normal setting. Many of the people who are in the shelter's program are . . . if they could do school in a normal way, they would have. They wouldn't be homeless. We have people with attention-deficit problems or recovering from, or not recovering from . . . sometimes people will show up drunk or whatever. Although that doesn't happen very often. They have to be sober to stay.

I really feel, though, that the thing that has happened is that we have created more community. The neighborhood has certainly been improved. The neighbors used to hate having the shelter there. Now they love the garden. The shelter residents know that they have skills. They know that they love working with plants, and they know that they can do that. They know about some health options that they didn't know about before. We learn a lot from them also. I am just amazed at some of the stuff . . . we are in lots of ways equal partners.

People are making community connections. There are homeless people who now know who grows herbs, and they know people from Tom's of Maine, which is a natural products company. They know some local CEOs [chief executive officers] and the president of the bank. They just know people they didn't know before. And those people know homeless people, which is pretty amazing. We have wealthy people and homeless people working together in the same garden. For all of its difficulties, it has been really pretty amazing.

FUTURE OF HERBALISM

We are in a really challenging time right now. What it will require to bridge these paradigms and to really create a whole healing system, is for all of us to really look at our own commitment, why we are doing what we are doing, what it is we love, why we have a passion for it. We are in a sense affirming our own commitment, and at the same time being willing to step back from it to really hear the other side, really hear the other person. I think that it is critical for herbalism to become acceptable and for us to find a viable healing system for this country. I really think that herbalism plays a critical piece or can play a critical piece in using our land sustainably and in looking at what the land really is. It isn't just some dimension written down on a piece of paper. It is an embodiment of something that we don't even really have words for probably, but obviously it speaks to us deeply. I think that when we have those intimate connections with the land we tend to treat it in a more sustainable fashion. When we know what is underneath our feet, we are a little more careful. It is like having a friend. I think all of that is important.

Chapter 20

Annie McCleary

Annie McCleary is a folk herbalist and owner of Purple Coneflower Herbals, which specializes in organically grown echinacea and goldenseal extracts. Through her business, Annie educates about the value of folkloric medicines and the need to cultivate at-risk medicinals. Spirituality and a deep attunement to the subtleties of the green world guide Annie at every step of her life. Over the years she has taught many people in central Vermont and beyond how to attune themselves to the spirits of the plant world through dowsing and other spiritual techniques. Annie also teaches the art of herbal preparation, plant identification, and wild food harvesting.

I asked Annie to speak about the place of Spirit in the modern herbal context, and how connection to a sense of Spirit affects many of the areas which may present challenges to the contemporary herbal community. We spoke on February 19, 1999, at her house in East Hardwick, Vermont.

INTENTION AND INFUSING SPIRIT INTO WORK

From the planting of the seed to all parts of production of herbal medicine, one needs to be present and prayerful. At each step, from planting to transplanting, to weeding, to harvesting and processing— it doesn't matter; it could be sweeping the floor; it could be washing up the dirty bottles. It doesn't matter the stage or the activity, the idea is to maintain a mindful stance, a grateful stance. These plants are giving to us of their incredible healing energies. This is a gift they're giving us.

When I'm mixing formula, I thank not only the plant spirit but also the soil and the wind, the sky and the air, the rain, all the people who helped along the way, my wonderful apprentices who helped plant and weed in the summer, all those helping hands who harvested the plants, people who helped me process. You know, it's a responsibility to be in the position of facilitator, moving the process along from seed to a bottle of tincture which will help someone in their healing process. There are prayers all the way through.

I'm a medicine maker and teacher. Herbal medicine involves much more than chemical constituents. It is not a mechanical process—simply grow a plant, chop it up, put it in—*no*. Herbal medicine making is full of intention, a constant honoring of the plant spirits. So by the time this little bottle gets to someone who needs it, this is charged; *this is charged*. A fellow came up to my booth at a local food and health fair. He told me that he uses my echinacea extract. Sometimes he ingests the extract, and sometimes he simply holds the bottle and feels a change in his vibrational energy.

Before I even step into the herb lab, I do my meditation. I smudge the house. I ask for guidance. It's a constant interplay, between me and the formulas and Spirit. It's not a rote situation at all—it's a very creative process. It's fun! It's a gift. I know it's my work. I have no doubt. I don't wonder, *Should I be doing something else?* I know this is what I should be doing, so I throw myself fully into it.

When I work with plants, I work with the vital energy that is there. I speak to the plants and give them my energy; they speak back to me and give me their energy. We're in communion; we're working together. They know that we replant. They know that the ecosystem is not being damaged by pesticides or herbicides. We grow our herbs organically. We're simply going with the harmony of the universe. If you want to restore harmony to the human body, you need to restore it with something that is itself in harmony. You can't restore harmony with something that's out of harmony.

Commercialization and Disconnection from Spirit

I think most herbalists, if they have any connection to the plants at all, have a sense of the spiritual connection. Native peoples in this culture and elsewhere have known and honored the spiritual component of healing. But as you know, in our current dominant culture, connecting healing with spirituality is not completely accepted. I really don't know what the bigger companies do, but I do know if they ignore the plant spirits, they are missing the mark. It is my hope and vision that people working with herbs open to the spiritual nature of plant medicine.

Does the Popularity of Herbs Alter Spirit?

To my understanding, Spirit doesn't change. Spirit is solid. Spirit is *the thing* that doesn't change. It's humans' connection to Spirit that is the issue, and it is a big issue right now, obviously. Look at what's going on on our planet. Are we maintaining a balance? Hardly! We're at the edge of disaster. But because it is a very difficult time, people are waking up too—you see, it's a double-edged sword. We've gotten ourselves into so much trouble not listening to Spirit. People are working hard, it seems, on their inner issues, using a great variety of healing modalities. For many, the process of healing has recently accelerated. I think this is part of the process of pulling us back from the brink, helping us through whatever transition we have to go through to reach a balance. It is an issue of humans remembering who we really are, and what it really means to have the gift of being here on the planet, and what that responsibility entails.

HOLISTIC VERSUS ALLOPATHIC PERSPECTIVES

I think we need to understand why Americans are flocking to traditional forms of healing. Would they be doing so if all their needs were met by the allopathic system? Healing involves all levels—physical, emotional, and spiritual. As people heal physically, they make emotional connections and heal on that level as well. And how often I see people with serious illnesses, such as cancer, grow and evolve spiritually, reaching inner peace and clarity. The person may even pass over, leaving behind the body that could no longer house the soul, yet much healing has been accomplished. Essentially, you see, we are spirits in bodies.

As people heal, they will seek modalities that address body, mind, and spirit to assist their evolution. Herbal and other traditional healing systems address the whole person.

HERBS AS A LIFESTYLE

I came to herbs out of the need for my own healing. I spent twenty-two years of my life teaching young children in public and private schools, rural, suburban, and inner-city schools . . . first grade, kindergarten, and preschool. I often had sick children in class, and I often came down with bronchitis and earaches. I realized I needed to improve my health. I also realized that the major medical profession knew little of prevention and wellness. So I started my journey, reading herb books and going to herb conferences and women's health conferences. I was, at the time, in upstate New York, and I studied with Susun Weed for three years—Tuesday and Thursday nights, psychic healing and herbal healing.

It is definitely a holistic, multidimensional process that I have been going through. Herbs are my allies and have played an important role in the healing I have experienced. Along with the herbs, I have made changes in my lifestyle, my diet, and my relationships to myself, others, nature, and Spirit. I learned how to heal my heart. I see how my emotions and my health are connected. As you see, healing is quite a process of evolution.

Echinacea has become one of my primary herbal allies. She has played a major role in my own healing, and I now make echinacea extracts—as well as extracts of other herbs—as my livelihood.

ORGANIC GROWING OF HERBS, WILDCRAFTING, AND THE LINK BETWEEN SPIRITUALITY AND THE RESPECTFUL USE OF PLANTS

One year, a farmer offered to germinate two hundred echinacea plants in his greenhouse for me, and I thought *Oh, my gosh, two hundred plants.* Later, we joined in a partnership of growing echinacea together, growing twenty-five hundred every year. Some of those plants I used in my own herbal business, and most were sold to other herbalists and individuals. Now, I'm back to growing two hundred or so echinacea for my herbal business. They're beautiful. And the goldenseal, too—we grow her organically here also. She's a bit of a challenge to grow. Goldenseal is a woodlands plant whose habitat is deciduous forest. We are growing goldenseal under shade cloth in the field.

I do some wildcrafting for my business. I wildcraft St. John's wort, yellow dock, burdock, and dandelion. I feel like I have excellent connections with them. Essentially, there's no difference between the relationship one has with a wildcrafted versus an organically grown plant. Both involve communion with the spirit of the plants.

I really got into *Rumex crispus,* yellow dock, last summer. Often, certain plants just sing out to me, and we deepen our connection. I started cultivating yellow dock because I didn't want to stress the wild populations on the farm. There's a huge yellow dock next to my compost pile. She's the mother, and you don't harvest the mother, you know.

I check in periodically on the status of osha since she's a plant that I use in my herbal extracts that I'm not growing or harvesting locally. I got that she was not to be harvested last year. My feeling is that habitat everywhere is being compromised by pollution and agricultural practices and development. We all know that. There is a lot of osha out there; some of it is on very protected land, but there may not be enough osha for everyone, everywhere. I will use osha only when I am sure the stands are strong and properly protected.[1]

Lack of Spirit and Overharvesting

Education is important. We must raise consciousness around proper wildcrafting practices. As one becomes more conscious of a plant's habitat, life cycle, and range, one is more likely to use the plant wisely. And we can use plants that are available to use in abundance, like dandelion and burdock. Not too long ago, folks knew the benefits of eating dandelion greens. My mother speaks of the person who brought around dandelion greens in the spring when she was a child in Pennsylvania. Hopefully, we'll come back to eating our dandelion greens! Burdock root is cultivated in the Orient for food, known there as *gobo*. Actually, it's cultivated here in central Vermont also. And we certainly need to encourage organic cultivation of herbs, especially such popular medicinals as echinacea and goldenseal.

HER APPRENTICES

For the last two or three summers, I've had young women come to me to ask if they could apprentice. I put out no flyers . . . nothing. I honor them for seeking me out. We meet one day a week from April to November. We start in the morning out in the garden, around the prayer pole, and offer our prayers. We do a check-in. We use a flower as a talking stick. Then, we are off to our work. We have a potluck lunch and go back to work in the afternoon.

My apprentices help with the work in exchange for learning about plants. There is no exchange of money. They learn gardening, cultivating and wildcrafting herbs, herbal preparations and more. They get a lot of work done. I'm fifty-one; I cannot get out there and haul and shovel the way I did when I was twenty-one. I'm getting calls already for next summer. I need help. And I want to pass on my knowledge.

The term *apprentice* is used differently by other herbalists right now. These apprentices are students who pay for classes—a wonderful form of learning. My apprenticeship is a revival of the older form—working in exchange for knowledge.

I also teach herbalism in the classroom format. In my herb classes, I like to focus on local plants for food and medicine. I teach how to identify, harvest, and prepare herbs, honoring the Spirit of the plants at each step. It's important to work with herbs and Earth in an intentional manner. In class, we *talk* about honoring the plant spirits; in the apprentice program, we *do* it. That's the difference. Both class work and the apprenticeship have value. They're both to be recommended.

RITUAL AND INFUSION OF SPIRIT IN DAILY LIFE

There's so much you can do to infuse Spirit into life. One of the simplest rituals that many people do before a meal is hold the bowl, hold hands, thank the cook, thank Spirit. It's simple.

When my apprentices go out and wildcraft herbs, I suggest to them that each snap of the stem reminds us to say thank you. [Whispers] "Thank you, thank you, thank you."

Infusing Spirit into life involves simply maintaining an attitude of intention, for the good of all beings. We are given a reason to be here on the planet. May that be fulfilled.

Chapter 21

Michael McGuffin

Michael McGuffin, along with Janet Zand, founded McZand Herbals, a national herbal product company, in 1979. Michael's innovative business skills and love for the natural world helped to make McZand a successful business venture. The company quickly put out its first three products: Insure Herbal, Female Formula, and Male Formula. More than twenty years later, McZand products are staples in health food stores across the country.

Currently, Michael is the president of the American Herbal Products Association (AHPA) and is a member of the FDA's Food Advisory Committee Working Group on Good Manufacturing Practices for Dietary Supplements. He is also on the executive board of United Plant Savers. In his work with AHPA, Michael is the media front man with quotes and appearances in *The New York Times, The Washington Post, US News and World Report,* as well as National Public Radio and ABC's *Nightline.*

I spoke with Michael over the phone on December 8, 1998.

HOLISTIC HEALING

There are a range of holistic concepts, and many people do not agree on what makes up holistic healing. The concept that I am most able to identify is something about taking care of yourself. This comes from my personal experience, my personal background, and my age and time and lifestyle choices. I thought it was interesting and a little in the face of the establishment to say that maybe if I were to learn enough about paying attention to my own body, I don't have to go to a professional. The professional need not be the best caretaker for the health of my body. Maybe I could play that role. That is clearly a holistic concept, self-responsibility.

Another holistic concept—it's not all that new—is balance and moderation. That expression is much older than the contemporary herbal movement, but it's a significantly holistic concept. I can drink a little wine several nights a week, but if I start drinking in the morning, I'm probably in trouble.

HOLISTIC HEALING IN THE MARKETPLACE

I think the primary function [of the herbal products industry], especially in the light of consumer self-medication, is to provide high-quality product and, as much as possible, to also provide meaningful, useful, not misleading information.

There are a handful of people who want knowledge and understanding [of holism]. There is probably a larger group who conceptually is at least intrigued, and in some cases really knowledgeable about the concept of holism, but a lot of the customers for herbal products are in for a quick, simple, just tell me how to do it, what pill should I take now because I don't want to take pharmaceutical pills. When I said we're trying to provide two things, quality products and information, that is the only possible way that we can communicate with the consumer about some kind of holistic concepts.

As for the lifestyle choices, I don't know if the herbal products industry (HPI) has a responsibility, an interest, in promoting what are really lifestyle choices. I mean, I could argue that we could, but then I could argue that we shouldn't. I could argue that all I want to do is provide a high-quality product so people can make these choices and not really attempt to affect their more basic and personal choices.

[The consumer of herbal products] is primarily asking for an alternative to a drug. They're asking for that for two primary reasons: that they believe and assume that herbs are generally more affordable, and they believe and assume that they'll be generally safer. There are other factors. There have been more studies on herbal products in the last fifteen years than there were in the century before. That information filters in.[1]

There are all kinds of cultural influences, too. I went to college in 1969. I think everyone who went to college in 1969 had the same major: countercultural experience. All we wanted to do was look at alternatives: alternatives to how we earned our living, alternatives to how we, as a nation, settled our differences with other countries, how we medicated ourselves, how we chose our food, how we wore our hair. Lifestyle choices are a segment, and a really visible segment, my generation made. To some degree we're still moving that out, and it's become more broadly acceptable. Many of the values that were attributed to the hippies are no longer fringe; they're perfectly acceptable, especially in the framework of our generation. Then, of course, there is the generation of our children who've been influenced by us, and even the generation of our parents. Our parents have started to deal with aging. They too are looking for a kinder, gentler approach to their health conditions. All the time, my parents want to know "What can I take for this?"

RESPONSIBILITY OF THE CONSUMER

The primary responsibility [of the consumers] is to inform themselves beyond the point where we've been able to inform them, or to have the requisite training to self-medicate. I don't believe that requires a doctorate in medicine. If I'm treating my cold or my headache, I think I have the training just in terms of my personal experience.

The other responsibility of the consumer is to recognize when their [health issue] is beyond their own knowledge and intelligence and experience. If I get six headaches a year, I can treat them with feverfew or ginkgo or kava. If I start getting recurrent headaches, I should have the good sense to go to somebody who has the training and experience to do a legitimate diagnosis. That's another subtle piece of the responsibility of the consumer.

[Education for consumers] is getting out there. It's hard to sort the good studies from the bad sometimes. It's not easy to research all the data. The Dietary Supplement Health and Education Act (DSHEA), for all its flaws, has opened the door to much better access to information. It clearly broadens the availability of information.[2] There's tons of information. Actually, I've heard somebody say "Lots of data but no information." That's always a possibility. I think there's both data and information. It does take persistence and intelligence to sort it out, and [the willingness] to suspend your own beliefs however. It is hard sometimes when I'm looking at abstracts of technical data. I recognize my tendency: I believe the ones that say the herbs are positive, and I assume the studies are flawed when the herb tested negative. I know that there are people starting from another perspective who have exactly the opposite response. The work, for me, is to try and maintain objectivity. I think that's true of many consumers. If you go into it and honestly believe that a natural substance is always better than a chemical substance, it doesn't matter how much information you're given, you are going to stay with that position. It is not just herbal medicine; it is delving into philosophy.

HERBS VERSUS HERBALISM

The herb is only the herb. The herb plus knowledge becomes medicine. The herb plus knowledge with somebody who shares your opinion becomes much more immediate and localized. I am not an herbalist. I am not a practitioner. I don't try to diagnose or prescribe anything to anybody but myself, but I really value those knowledgeable individuals in the community—teachers like Deb [Soule], Pam [Montgomery], Chris Hobbs, Roy Upton, and David [Hoffmann].

The Herbal Product Industry (HPI) makes six or eight billion dollars, and what part of that goes through the hands [of the herbal practitioner]? Not much. What part of it goes through retail setting? Almost all of it. What part of that part in the retail setting is touched by a deep knowledge [of heart-centered, Earth-centered] herbalism? Some, not much. So I think that what's being marketed is herbs—herbal products, herbal formulas. Nobody's marketing herbalism [and connection to Earth], except the individual herbalists who agreed that they would publicize their activity in their community.

You can't sell people something they don't want. If the consumers wanted [the spirituality of herbalism], they would have told us they wanted it a long time ago. I don't think they want it. They want a suggestion of it; they want the icon. It's been interesting to me that almost all of the herb companies have a practitioner as a representative of their brand. So Janet Zand, who's our formulator, is a trained herbalist and acupuncturist, and Christopher Hobbs worked with Rainbow Light for years. Dr. Christopher really was the first one, in our time, to be associated with a company, Nature's Way. Rick Scalzo is an identity for Gaia. Rosemary [Gladstar] now works with Frontier. There is some suggestion that the consumer at least wants to know that the herbalist was involved.

I don't see any interest in creating and marketing herbalism, as I understand it, which has a lot to do with communication between a knowledgeable practitioner and a patient. But again, then we get into this thing about, "How badly do we need the experts?" There is not a fixed answer. Do you need a degree or outside expert to medicate yourself more than you do in order to feed yourself? Maybe you do. Certainly in extreme instances you do. I am not an advocate of saying, "If you find out your sister has leukemia, you should give her a lot of wheat grass and do research on the Internet." No, no, no. That is not a position I would take. At the same time, it starts in a sense to go back to the conversation about responsibility. I've got a cold right now. I'll manage it by not going out of town, even though I already had the tickets in my hand, and just taking a relaxing bath. But I can still come to work. I am taking it easy and taking my herbal medicines. Then again, sometimes I want to go out and bring back a higher level of expertise. When I came home from Bali with dysentery, I called both my acupuncturist and my MD. I treated myself, but I did it under their supervision.

PROPER USE OF HERBAL PRODUCTS

I'll go ahead and admit my bias is, "Don't treat your cancer herbally." I know a young, personal friend. She was twenty-nine years old and got leukemia. She went through this whole self-examination about how she would treat that. She made her decision that she would treat it with chemotherapy. She called all of her friends together, and she said, "Here's my decision, and I need to make this clear to you. I love you all. I'll be friends with you all always. Everybody who doesn't want to support me in this decision please leave me alone until I've recovered. Don't tell me that I'm doing it wrong. I don't want to hear about it. I can't hear about it. I'm trying to save my life." That woman is still alive, and she's quite healthy. She had this horrendous eighteen to twenty-four months. I know two people who didn't do that and died. In one case, I think a really irresponsible decision was made. I don't want to do anything to support that irresponsible decision. So I don't want to say chaparral is a liver cancer cure. I really don't. I don't even want to make it available to a well-informed practitioner who believes that it's a liver cancer cure. I am satisfied using herbalism to treat self-limiting conditions, and also as a support in long-term conditions. I think herbal medicine can be a great support if I have a family history of a digestive disorder. Then it really makes sense for me to use herbs as a long-term preventative and control. The same thing if my family had a history of heart conditions or circulatory conditions. That's another part of the responsibility of the industry: to recognize that there are limitations to what we ought to be self-treating. If you look at other models (especially the European and maybe even the Canadian), the natural remedies all have been regulated. The regulations exist to allow herbal remedies for certain conditions only, and they all are self-limiting conditions. They are specific, and aren't potentially fatal conditions.[3]

I think that herbal medicines or dietary issues or ideas can be a significant, supportive part of the healing in one of the more fatal conditions, or potentially fatal conditions. Then I think too, that the right thing to do is to bring in some expertise. Don't think that you can learn it all yourself in a couple of afternoons on the Internet. It is a balance.

At the same time, I don't want to discard the drive and intelligence of the individual. I know myself, I can start knowing nothing, and if I make it my path to know something about something, give me thirty or sixty days, and I can build a fairly considerable bank of knowledge. My tendency would be to go test that out against somebody with more expertise than I can get in writing, but there's something about finding a balance point [between] assuming that all of the expertise resides in the expert out there and assuming that I can learn anything. There is a flaw in both of those, and there's a middle ground that's probably the best place to take in terms of finding those supportive therapies.

ENDANGERED PLANTS

I have contracts for thirty-five acres of goldenseal. It was a strict, simple, business decision. A strict, simple, supply-and-demand decision. What I saw was my business growing potentially out of my control. As a businessperson, that's unacceptable. The only way to run a successful business is to control as many elements as possible. (Of course, everything is never in control.) I started seeing a potential solution long before I started seeing a problem. Clearly, with regard to supply, that is the place of the most significant concern. I wasn't that worried about the plants from which leaves are harvested. I'm still not. I was more concerned with those that are harvested for their roots and bark. I wasn't all that worried about sassafras, however, because even though you take the bark and the root, it's a weed. I was worried about goldenseal. I was worried about slippery elm. I was worried about echinacea.

Your question is, "What can the industry do?" Well, it varies on different products; it varies on different companies. What we did is we stopped using slippery elm and converted the slippery elm in all of our formulas to marshmallow root because we determined that was a legitimate replacement. I don't think that's the appropriate response for a company whose entire business is identified with slippery elm. Nor do I think the appropriate response is for that company to hold a press conference to announce that they are going to go out of business, that they are going to stop selling slippery elm because it may be in decline. I think the only appropriate response from such a company would be to plant some slippery elm or start reformulating. I think that it is important to recognize that there are so many different perspectives. The simplest way to state the business perspective is that the first response to a supply-demand imbalance must be an increase in supply, not a decrease in demand. I am not interested, from a strictly business point of view, in running and saying to people, "Please stop buying herbal products because maybe there's not enough herbs. We'll get back to you when we think there are." I think that would be bad for my company, bad for my employees, bad for the industry.

So what can industry do? It gets to be very specific. I can only tell you what my company did. My company has invested close to three hundred thousand dollars the last four or five years planting *Echinacea angustifolia,* because that is the species we want to use, and goldenseal. We lost on echinacea. I know what it feels like to plow dollar bills into North Dakota dirt. It's kind of fun, but not for long. But we kept trying, and actually we did get thirty-four hundred pounds dried echinacea root through two different farms that we contracted three years ago for echinacea. That came in the day before yesterday.

I started harvesting cultivated goldenseal in 1994. We were planting it right around 1990. We have not bought any wild goldenseal in the last two years. We probably will again; we probably have one more year where we are going to buy some more wild goldenseal, but we're moving in the direction of converting it all to a cultivated source.

As part of my work through the American Herbal Products Association, of which I've been a board member for eight years, I am the point person for the communication between the trade association and the U.S. Fish and Wildlife Service. So we are coordinating an actual tonnage survey to quantify the amount of goldenseal that is being harvested wild and the amount that is being grown.[4] We are going to be comparing that to historical data so we know what, as an industry, we are doing. It is early. The preliminary data look pretty good. I have got data that's one hundred years old from five different sources that suggests that, historically, the yield of goldenseal has been about five percent from agricultural sources. The first information that I looked at, on the current numbers, suggested that it's closer to twenty or twenty-five percent right now, or will be within two years.

MONOPOPULARIZING

With regard to monopopularizing, the herbal product industry doesn't do that. The media does that and then we try to keep up. We don't go out there and say, "The only thing you ever needed for anything having to do with the female reproductive system is black cohosh. Please run it on the seven and eleven o'clock news." The media picks up a little story and they run with it, and yes, then we try to keep up. Then we do try to meet those sales. We didn't decide not to. We didn't look at it and say, "This is a problem. This is a small plant. We don't know what the supplies are." We simply haven't caught the wave yet. I don't know if there are any plans in the marketing department to do that. I know that if I am aware of that I would certainly push all of the soy research on them that I could. But I know what they would they would say: "Yeah, but *USA Today* ain't selling soy. *USA Today* is selling black cohosh." That's a tough thing to ask a company—I think you should keep asking it. It's a good challenge.

If you are aware of a monopopularized herb, of an herb that's suddenly in the spotlight, and you don't have that product in your mix, one: Do you have to add it? There might already be a half a dozen good products, and two: Do you have to add it before you examine the environmental impact of that much more of the product? That's a good question. I am not really prepared to answer that for the industry, and I am barely able to answer it for my own company. I will acknowledge that's a good question. But hey, let me back it up fifteen years and, as far as I'm concerned, nobody else needs to have an echinacea-goldenseal product. We have been doing that for a long time. The other herbal product companies should just put out empty bottles and say, "see Zand." But they aren't going to do that.

The other part that you really have to remember is, to some degree, this is just a business like any other business, and while you will find some ideology and some sense of higher good, possibly more pronounced in the industry than in others, there still is a real tendency to just push forward with providing the consumers what consumers are asking for and forget all those other issues. I am not saying that it should be. I am just saying that's how it is. We are a business just like any other business.

I do not think there is any good information to show that goldenseal is in imminent danger of significant decline. The report [WWF, 1999] that Traffic North America put out was flawed. They knew it was flawed. I am unimpressed with the fact that goldenseal is rare in Alabama and Mississippi, because guess what—it was rare when Bartram visited Alabama and Mississippi in 1763. I am unimpressed. They carried that document in there with an intention to get a listing, and they didn't ask for a peer review of their document. They didn't ask for anybody else's opinion. At the same time, it is fine that it exists that way. It kind of turned the lights on. But, I will say that I don't believe that's an effective long-term approach; I do not believe that you're supposed to initiate a conversation by excluding a primary party, that is, industry. I don't think that is in the common good. I don't think that can give you the best outcome. And I really don't think that goldenseal is in trouble. The black cohosh maybe, but I don't think the goldenseal is. But I've made a business decision. I can control my cost in supply if I plant it out in straight rows. So I do.

FOLKLORE IN THE MARKET SETTING

The DSHEA established a definition of a new dietary ingredient, and specifically acknowledged that any dietary ingredient that was already marketed should be considered as a safe dietary ingredient. In effect, what they did is grandfathered in all of the traditional information so products that were already in the marketplace in the United States were okay with regard to their legality of sale. With regard to what you can say about them, it's a whole different thing. Statements of nutritional support require that the statement be substantiated. It doesn't say "substantiated with double-blind clinical studies." It says "substantiated," which I believe leaves a whole lot of room for a manufacturer/marketer to determine what kind of substantiation depending on just where you want to make your stand. The law says I can make a statement of nutritional support as long as I can substantiate it, and I am safe so long as my substantiation is in order prior to my making the claim.

The way that my company does it is we put files together. If I am willing to put together a file that says that black cohosh—I'm going to go ahead and make a statement of nutritional support—is a parturient, so black cohosh based on Murray's work, or his republication of an earlier herbalist's work from 1760, my insurance people are going to shudder. They are going to say, "Please, Michael. I don't like that degree of substantiation. Couldn't you find some twentieth-century clinical study?" So in a sense, where I think the law would allow the use of folkloric sources of information, my business decisions are not going to allow me to call on that data unless it is really well documented, kind of redundant and broadly reported to stand. When I put these files together, my most thorough one says, historical data, traditional use, and contemporary study. That's what I call substantiation. But if I can only get one [category of source material], and I know that the second one is the one that will bear up to scrutiny more, I have a higher degree of restfulness if I've got the second. It is strictly just a business decision again. The FDA isn't essentially outlawing traditional uses, but they're not making any room for them.

Chapter 22

David Milbradt

David Milbradt is a state-licensed certified acupuncturist and a professional member of the American Herbalists Guild. He has studied with 7Song, the director of the North East School of Botanical Medicine, as well as nationally renowned author, Matthew Wood, and many other prominent teachers of herbalism. David completed an intense herbal internship at the Center for Alternative Living Medicine (CALM), a mobile "hospital" which sets up in wilderness areas across the country.

David has taught herbalism at the University of Wisconsin-Madison minicourses since 1993 and is also a founding member of the Madison Herbal Institute.

I spoke with David over the phone on September 30, 1999.

DEFINITION OF HERBALISM

I would say that herbalism is a modality for influencing health and disease, mediated mostly by plant substances, although certain minerals and the like could be included. I don't use them a whole lot, but I would include that. That would be my answer, using herbs to influence disease.

ROLE OF THE HERBALIST

There is a tradition of saying that the herbalist is not a healer. [He or she] simply helps the body to heal itself. It seems like dodging the bullet to me [laughs]. I would say the herbalist is one who has expertise in this modality of healing. He or she has a calling for it, hopefully. I would call someone an herbalist even if he or she didn't have a technical proficiency. I think an herbalist is a doctor trying to affect health and disease.

LICENSURE FOR HERBALISTS

I am licensed as an acupuncturist. That has a standard of practice in Wisconsin that allows us to practice herbs legally. It is part of the standard of practice. Right here in Madison, there are a number of naturopaths who are doctors of naturopathy, whether they got it through a correspondence course or not. At whatever level of studies they got their doctor of naturopathy, they are not licensed to practice medicine. They advertise in the yellow pages, which I find amazing. The level of insecurity in building a practice to have one angry client easily destroy it by just causing legal problems—I found that level of insecurity to be difficult to consider before I was certified as an acupuncturist.

In thinking about licensure, there are a couple of aspects to consider. One is whether it is practical to do this, to license herbalists, and still have any depth to what they are practicing. For example, in England, there is a common law that allows herbalists to practice without being licensed. This means that you can hang your shingle as an herbalist without certification, and the law can't mess with you. If you hurt somebody, I am sure there are repercussions, but you are not dangling your butt in the wind like these people here who are practicing and advertising in the yellow pages. I would imagine that in England they could have malpractice insurance. And they cannot be shut down for practicing legally. Now that is a pretty ideal situation.[1]

I don't think licensure brings competence as is commonly thought. What it does is allow a group of people to professionalize. Then maybe they can raise their standard of living as a profession. Then they are monopolizing the area they are in. They are restricting access for other people. That gives them a space where they can create a financial base. There is a real positive aspect to that. People should be paid for doing what they do. But there is also the restrictive aspect of it. The chiropractors had a huge battle with the AMA, a running battle that lasted fifty years or so, before they were recognized and professionalized [Whorton, 2002, pp. 178-182]. Now they are legal and have insurance coverage. Here in Wisconsin there was a type of chiropractic . . . the name is slipping from me, but they did a very gentle manipulation. They claimed to feel the effect of spirit to some extent. This is not really wild in terms of chiropractic if you go back in history. Anyway, the chiropractic association here in Wisconsin made that practice illegal through a change not in the legislature, but a change in the committee that regulates chiropractors. They made it illegal for [people] to advertise that they did this type of chiropractic and to even speak about it. So, here we have a group that was persecuted for being outlandish, persecuting themselves and drawing restriction on what can be done and what can't be done in the name of chiropractic. So, professionalism creates bureaucracies and bureaucracies create definitions. As they get older, bureaucracies get less and less rational.

Now, herbalism in America is as deep as it is wide. Well, actually it is wider than it is deep [laughs]. We could use some depth of tradition. It is not our fault that our roots have been scorched and tortured and cut off and transplanted and retransplanted again. But there it is. We are trying to get our roots back in the ground again, and see where we are.

This is a society that recognizes highly defined standards of practice. To contrast herbalism with Chinese medicine, Chinese medicine is thousands of years old and has a tremendous amount of diversity. However, in the 1950s, late 1950s, Chinese communists decided they were going to create a state brand of acupuncture and Chinese medicine. Committees got together and they defined what the state brand of Chinese medicine would be. That was then taught in schools. Amazingly enough, this was all consolidated before it was imported over here to America. That gave Chinese medicine an incredibly unified standard that it could not have had twenty years earlier.[2]

Within Chinese medicine, there is TCM, traditional Chinese medicine, which would be better described as textbook Chinese medicine. That is the unified front. There are other schools in the West, because [J. R.] Worsley came over from England. The Worsley schools claim to be really traditional [laughs]. It is a bizarre type of wording. Their information came from pre-TCM Chinese medicine. So, they have one perspective of the big elephant. Everybody here has to take the same TCM test to be nationally certified, regardless of what school you study in. My school emphasized traditional Chinese medicine. Other schools emphasize the spiritual and emotion aspects, which, of course, weren't the biggest issues that Chinese communism was trying to get across. They were confronted with Western medicine and were trying to look as scientific as possible. Then there are also communist ideas about spirituality and materialism. If I were in a school like that, in order to study the boards, I would have had to crack the books, and have a more clear picture of TCM, and how to approach the stuff from that point of view. I studied TCM so now I have to crack the books to explore the more spiritual, emotional aspects of Chinese medicine. They are there. There is no way you can hide.

If herbalists are going to have national certification, there will need to be an exam. I don't think it can be done without that aspect. That will have to have some agreed upon, basic fundamentals that you need to take that exam. All these certification programs mean is that you have the capacity to undertake certain academic challenges. That is as far as it goes. It also means, probably, that you can go through a certain amount of school, take the test and a few other minor requirements.

Each school sets its own standard in terms of how much time is involved before you can graduate from the school. I am a certified acupuncturist. I did not study traditional Chinese herbal medicine at my school. So, I have never taken the herbal boards. But that doesn't make a difference in Wisconsin. Herbalism is part of the standard of practice as it is set up. The training is not even defined.

SCIENCE AND FOLKLORE IN HERBALISM

Sometimes I see myself trying to learn the details of the craft and convince people that those sort of scientific proofs are there, the validity, the rational proof as it were . . . and at the same time trying to hold onto the real proof, which I feel is based in the tradition and spiritual connection that I have.

Simon Mills [1991], in his book *Out of the Earth: The Essential Book of Herbal Medicine,* says that one of the differences between the doctors and the herbalists is that the herbalists have better stories. This is important because if you can't help a person to understand how [he or she] can move from disease toward health, if you can't give a clear picture of how that can happen, the patient may not be able to visualize the step he [or she] needs to take to get there. The scientific story of how the body is making that step can be every bit as important and operable as another story. In some ways, science gives us a tool there to help explain to the person how [he or she] can move from disease toward health. In that sense, it is very useful.

Science can offer certain proofs, at times, of the validity of certain herbs. But with the science, you often get into what Ed Smith calls blackboard science. That is where you sort of jump to conclusions that because St. John's wort is an MAO [monamine oxidase] inhibitor, and MAO inhibitors interact with anesthesia to cause a spike in the blood pressure, that St. John's wort will cause that same effect. This is blackboard science. Science puts the pieces together in such a way that it can lead to a conclusion that can be really far off base.

Take the example in nutrition. It is so close to herbalism because it still has the same nebulous polychemistry that herbalism has. This whole thing about combining proteins is a great example. Science shows that amino acids are needed to make proteins and that the human body doesn't manufacture certain amino acids which are called "essential." In order to have a complete protein you need to combine beans and rice for example, because the mixture supplies all of the essential amino acids. People are really invested in this idea. They will look into other cultures and say, oh look, this culture combined beans and rice and this culture did the same thing. It turns out that the science that started this idea was based on rats and rats have more essential amino acids than humans do. For humans, potatoes are a complete protein. Rice is a complete protein. Assumptions are made on these little bits and pieces in a way that traditional cultures don't do. That can lead you really far astray from practical reality. How did the Irish survive?[3]

There are two specific issues here. One is that science, unlike traditional knowledge, makes advances by disproving the knowledge that is current. That is different from traditional societies, which build and integrate. And compile and combine. The other issue is there is no validity given to the ancient in a scientific framework. Because science is so rapidly changing, it becomes very complex until the point where a specialist can't know his own area. Then, it has to be broken apart into even smaller pieces.

One of the kernels that I like to chew on is this idea that one of the differences between scientific and traditional modes of thought is that in science unless something is repeatable, it is insignificant. A double-blind study is perhaps the best example. In traditional societies, events that are unique to the point of being completely and utterly unrepeatable, something happens which can never happen again, become more and more significant the less repeatable they are.

PUSH FOR CLINICAL TRIALS ON HERBS

The *German Commission E Monographs* [Blumenthal et al., 1998] is a book that touts itself as being a compendium of what has been proven about herbs so far. There is a lot of blackboard science in it. There are some things, however, that have been proven in terms of studies, clinically tested, and useful information about herbs there. But as an herbalist when I finally got a copy of it and had a look at it and poured over it, it felt like it was toolbox with only a few of the tools left. You have got a screwdriver, but what happened to the saw and the needle-nose pliers?

The tools are a diversity of herbs. If you are only going to have monographs on those that have been proven, you can only prove so many. It is expensive to prove. Then you have some herbs where you have only proven one aspect of this herb, only one action. It is a very limited way of using herbs. Generally they are used singly, not in combination. That is probably the best way for doctors and those in conventional Western medicine. It tries to be academic, academically proven. As a practicing herbalist, I am used to the gray areas, the thousands of different constituents that are in a constellation of utility. You have herbs that are like Swiss army knives. Elecampane, for example. Volatile oils, antibacterial, demulcents that are soothing to the digestive tract, resins that are slightly irritating to the respiratory tract which by reflex promotes expectoration [Mills, 1991, pp. 476-478]. All these things in one herb. With the proper conditions, all those different aspects of the Swiss army knife could be used all at once.

This is the primary difference between an herb and a drug. Bacteria continue to learn to resist antibiotics at a geometric rate. Bacteria learn to resist antibiotics more easily because medical science is looking at one constituent at a time instead of a multitude of different constituents. It is like echinacea. Yes, it is antibacterial, but it also stimulates the immune system, and also increases the activity of hyaluronic acid [Newall, Anderson, and Phillipson, 1996] that helps to form collagen and connective tissue. It is doing all of these things at once. It is a harder target or a harder weapon to work around. It is also something that is integrated into the environment. It doesn't put undo pressure and stress on the bacteria the way antibiotics do. So, that doesn't push the bacteria into situations like *E. coli,* a harmless bacteria which has turned very nasty. The fact that there is a multitude of constituents, a synergy, is a really big issue but also, from my point of view, being able to look at things from a traditional perspective is really necessary.

ALLOPATHIC VERSUS HOLISTIC HERBALISM

Western herbalists, many of them, practice in an allopathic manner. If you have a symptom like a cough, coltsfoot helps reduce coughs—give them coltsfoot. Yes, coltsfoot has more than one constituent and works in a more holistic fashion. It doesn't have a lot of side effects, but it is still referring to the allopathic way of approaching things. We do this primarily because our tradition has been diced, burned, cut off . . .

There have been some fairly complex, but generally very simple, straightforward approaches to using herbs. Things that aren't forgotten when they work. Chew a plantain leaf; put it on a bee sting. Working with herbs in a more constitutional, broadly balancing way requires diagnostics. That requires a framework of a tradition that has been pretty beat on. In Chinese medicine there is definitely that framework there. I wouldn't say that I have a complete grasp of Chinese medicine. I just recently started practicing as an acupuncturist legally. There are, even at my level, diagnostics that can be very helpful in dealing with the whole picture, the underlying root causes of patterns of imbalance in a person through that framework.

Chapter 23

Orest Pelechaty

Orest Pelechaty is a nationally certified acupuncturist and Chinese herbologist. He was the cofounder of the Aware of Life Options and Healing Arts (ALOHA) clinic, the largest holistic health clinic in New Jersey. Orest recently retired as the clinical director in order to pursue his own clinical practice and spend more time teaching and consulting with other health professionals. Orest works throughout the country and the world in the holistic health care arena. He has completed postdoctoral work in Korea and founded a holistic clinic in Tso Pema, India.

Although Orest spends much of his time in the clinical setting, this interview revealed a strong political voice, one that may help to bridge the gap between allopathic and holistic medicine.

The interview was conducted on September 3, 1999.

AMERICAN HEALTH CARE AND THE PLACE OF HERBALISM WITHIN IT

Basically, what we are dealing with is a role or a niche for a health professional that has not really been adequately met in contemporary society, certainly not post–World War II, by any means, but not after the Flexnor Report . . . since the early 1900s. Herbalism as a form of professional medical intervention virtually went extinct. It has not ever fully recovered. What we are looking at unfortunately in the West is a long tradition of political and economic suppression of folk healing, of indigenous traditions, of individual empowerment, of community-based, and family-based approaches to getting things done. We are constantly being pushed toward giving our power away. Big government knows better. Scientists know better. Pharmaceutical companies will take care of it. To the point that even reality itself is now defined through lenses that are controlled by multinationals. The research that a physician or a health care provider reads is funded and, to a large extent, influenced, and sometimes actually tainted, by inappropriate financial, economic, political influences, mostly pharmaceutical companies and petrochemical companies. I think it is interesting to see that New Jersey has the largest concentration of pharmaceutical houses.[1] [New Jersey] was also the first state to have the major refineries and the major oil companies as corporate entities. I think that there is a real connection there historically.

What we are looking at is taking what has often been a folk or Indian or barefoot doctor thing and bringing it into a respected role in an institutionalized, government-regulated, culturally constrained niche of a health profession. I see that as almost an insurmountable task. It is really very challenging. What is happening, in lieu of having an integrated role for a properly trained health professional, a medical herbalist, is that we are dealing with a tremendous interest and well-grounded intuitive understanding that the petrochemically based, chemical prescription, reductionist thing is not working. Plants are part of our heritage. Plants have been used in all cultures, forever, as a form of healing in a gentler, safer, more sane, and humane manner. People have this attraction to plants.

Unfortunately, the information out there is inadequate. Also, there is a range of products that for the large part are inactive, tainted, and of inappropriate quality. People don't really know what they are doing in terms of self-medicating. There is a tradition of doing it yourself, but that was for people who lived in communities who had extended families, clans structures, that had the grandmother to tell you. We don't have that. We don't have the normal safety parameters. We don't have a niche for a medical herbalist as a professional, but we have this tremendous need, this tremendous yearning. Now there is industry. The herbal industry is providing materials so people are going to be taking stuff with no idea of what they are doing. Or, sometimes only a partial idea of what they are doing which is worse than no idea, because then they are going to have some false sense of security. There is no guidance really.

Until a few years ago MDs would just laugh at people, actively dissuade them from doing anything herbal. "You are going to poison yourself." "That's quackery." "It is placebo." At least now, they understand to some extent that it does work. Now the opposite thing is happening. The medical professionals . . . at this point it is kind of schizoid . . . a lot of them are still saying, "This is all BS. Avoid it." This is knee-jerk, bizarre bigotry against botanicals. At the other end [are] the more forward-thinking physicians who are being squeezed by the insurance crisis. They are saying, "I need another source of income. My patients want this stuff. I will give it to them." Now they are jumping into it with virtually no idea of what they are doing.

ROLE OF THE HERBALIST

It is difficult for me to say what an American herbalist is. There are a few of them, and they are all a little bit different from each other. They don't fit a clear-cut sample yet. That is the thing. In the American Herbalists Guild, you have a range of people—lay herbalists, nurses, midwives, acupuncturists, maybe a few physicians, a few chiropractors. Outside of naturopaths and Oriental medical practitioners, there are no herbalists except in this very spontaneous, free-form, we-are-not-sure-yet-what-it-all-is renaissance. The American Herbalists Guild is a phenomenally important agent for the evolution of this health provider role. Unfortunately, it is not defined yet.

What I think we should look at are models from different places that have worked. The two contemporary societies that I think have something that works are Britain and China. In Britain, there are medical herbalists. They have professional training; they have a professional association [National Institute of Medical Herbalists]. They have standards. They have a materia medica, pharmacopoeia that they work with. They have been in a healthy relationship with their government regulators. The British herbalist is a freestanding health consultant. They are not doctors, but they are allowed to give guidance and to dispense materials. But in a very strange way, even there it is weird. You can't have the herbalist dispensing standardized manufactured products. They can give materials that they themselves procured or grew, but they cannot label them. They give this material to the patient, to the client, the customer. "This is what it is, this is how you use it." The prejudice is that it should be a tea. It is a gray area for them if they can use tinctures or not. Even if they do, they have to be self-generated. They cannot be from the store. That, I think, is the primary example we should be looking at.

In China, we have the barefoot doctors. Post-Maoist revolution, there was a tremendous need to get health care to the populace. What they did is they integrated a lot of the folk medicine into a basic training. They then plugged that into their public health system, and delivered basic care out into the population. You have people who are paraprofessionals. They are like paramedics who would do herbs and health screenings, appropriate public works, like disposing of waste. Basic stuff. That gives a possible clue as to what we can do in this country in terms of some complementary medicine, physician assistant role. That I think is a valuable avenue to pursue.

How can other herbalists fit in? I see a couple of successful cases. There are nurses who have gone through serious herbal training. They have this medical background from Western licensed accredited training. I certainly see an important part of it to have allied health professionals get a serious training in herbalism, not a cook book, fly by the seat of your pants kind of seminar approach. I live in terror of that now that there is an herbal *PDR* [*PDR,* 2000] that MDs, in their typical medical deity hubris, are going to crack open the book and go, "Oh look, antidepressant, St. John's wort," and think they know what the hell they are doing when they have no idea. They don't have enough training to do that. Much less to give any kind of due to the more subtle, the more holistic, the more traditional, and, I believe, the more true applications of herbal therapies. They are missing a hell of a lot. I don't want to see that become the norm. That is ridiculous. That is probably worse than people self-prescribing because quite frankly a lot of people will educate themselves fairly well. A lot don't, but I have seen a lot of patients that come in well informed. They are doing their homework. They are reading. They know a hell of a lot more than even an interested MD, in terms of herbs.

I think there are two ways for herbalists to go. One is to have some integration into the existing legislative regulatory structure so that they can be an allied health professional and an herbalist. That will be great because MDs don't have time to learn herbalism. They don't really have an interest or inclination to learn it the way it should be learned, anyway. They don't have the intuitive component, the more holistic approach, the more comprehensive approach that is part of appropriate herbal healing. I really think it should be done by allied health professionals, not by physicians.

There may be the use of herbalists as educators as well. They could work in a seminar format. That may be the role for the lay herbalist. They could be the health educator, one-on-one, or health educator in a public classroom setting. That is useful. However, then the question is, How does that plug into other health professions? If you are just a health educator, that's okay, but I don't really think you can do herbal medicine at a really high level in that model. You can handle the basics which handles most human needs, but, then there is a question of how do you really plug more coherently into a health provider team. That unfortunately, at this point, can only be through a person going through a more conventional medical profession and then bringing in the herbs.

HIS OWN EXPERIENCE DEVELOPING A PRACTICE OF HEALTH CARE

There are multiple challenges as to how to understand this, how to take it from here to a new plateau. Clearly, there is a need. The question then is how do we meet that need. It is a very much open-ended discussion at this point. What I can speak about intelligently is my own experience. For me to work as a Western herbalist, I have had to take into account a number of things, the most important one being legal status. Since my recommending herbs to anyone, besides as a health educator (and that it is a very weak shield), is illegal, I went into Oriental medicine—acupuncture, which is what is licensed in my state. We don't have a doctor of oriental medicine degree yet, but we have an acupuncture licensure that allows some degree of herbal recommendation.

I use Western herbs in conjunction with Oriental medicine. That has raised its own challenges as to how to integrate them, when to use what, and so on. That is a whole other area. How will Western herbology fit into Oriental medicine? How will that fit into the new world medicine, which is clearly evolving as we enter the new millennium? That is a whole other scenario. How does all that synergize, and get synthesized? Where is that going to evolve? In my case, I use western botanicals as an adjunct to . . . as a component of an eclectic medical approach rooted in Oriental medicine as a philosophy, under the licensure for Oriental medicine, or, in this case, acupuncture as it is in my home state.

PARADIGM SHIFT IN THE MAINSTREAM AND THE POTENTIAL OF HERBALISM

I don't think herbal medicine is influencing a paradigm shift in the mainstream. I think it is driving it. It is one of the engines. We have the vitamin and supplements area; that is one wave. I think another wave is herbal medicine, but again, it is self-prescribing. It is working through the mass media, the fad of the month, the herb of the month, but it is still getting stuff out there. People are thinking it through. That is actually a driving force for that decentralization of medical authority. That is absolutely happening and is one of the gestalts that we need to be looking at as we enter the new millennium. The pyramid structure with the medical deity on top and their underlings—the nurse, the physician assistant—eventually it peters down to the poor patient on the bottom. That could potentially be inverted. An ideal situation, of course, would be that people would be self-responsible, and that is a big, big step. It is quite a difference between someone reading a magazine article saying, "I heard that I can take this instead of this drug," and that person really understanding the deeper philosophical implication, the deeper systems approach to what they need to be doing for themselves. Yes, I think this is where we can have the messiah of the new age. The herbalist as prophet.

You can get into the green stuff, plants, which people are terrified of and look at it, smell it, touch, identify with a field guide, do the left brain but also do the right brain. Be intuitive with it. Learn what grows around you. Pick your own plants. That is a tremendously powerful healing for the planet, for people to understand, "Oh, I can't pick this plant because there are pesticides being sprayed around here," or there is DDT, or whatever they are spraying now for mosquitoes. That is an incredible opportunity for people to reintegrate into their biosphere.

Unfortunately, as useful as the Internet is—and I do believe that electronic technology is a major revolution probably equal to, maybe surpassing, Gutenberg and the printing press—as important as that is, it also is a two-edged sword where it is further alienating us from our bodies, alienating us from communities, alienating us from our environment. It is putting people more into their heads. It is just eyeballs and a brain plugged into the Internet rather than a whole person going out and looking at what is in his or her yard.

From my background, whether it is Chinese herbal medicine or Tibetan medicine, which is a personal area of interest for me that I have been studying for a long time, or Western herbalism, we know that it is good to have a relationship with your ecology, whether it is the ecology of your body, or your body in an environment. These are important parts of health and healing. It would be really good for people to go and learn what is going on with their food chain, and how they can use plants around them. I don't want to go too far afield here, but I think it is remarkable that a lot of herbalists will tell you, especially if they feel okay talking about the intuitive and nonlinear level of reality, that plants come to people. There is no coincidence that dandelion is one of the most common "weeds" throughout America. We know that as a medicinal, it is a liver detoxifier. It is difficult to find a patient who doesn't need dandelion or could not utilize it as part of their health regimen. The fact that it is ubiquitous says to the herbalist this is Mother Earth trying to send us something helpful. Just to even hear that, not even to believe it or experience it for yourself, but just to understand that as a concept, is a major step in consciousness raising. One of the powerful things about Western herbalism is getting out there, getting your hands in the dirt and planting some stuff even if it is a window box in the city. Getting in touch with life cycles and seasons and relationship with the green nations. Tremendously powerful stuff. We are looking at a paradigm shift and a real raising of consciousness. That it is an area that the herbalist as educator could really help with.

LICENSURE AND THE POTENTIAL "SOFTENING" OF SCIENCE

You can't license for the intuitive aspects of herbalism. It is impossible. You can't even do very well in terms of educating on a standard level. It doesn't work well. However, I am a big optimist that what we have currently considered to be science or rather the perversion of science into reductionist materialism is not something that will stand forever. The scientific approach, which is to test hypotheses, has for a couple of centuries been tremendously skewed toward a left-brained, male-dominated, patriarchal, abstract, misogynistic construct. It has been polluted by that. It doesn't intrinsically need to be that way. You can use logic and reason and experimentation to approach other phenomena that have been excluded a priori from this skewed, stunted view of reality. Science can indeed be expanded to include larger paradigms, nonmaterialist paradigms. There is that possibility. There could be a renaissance within science. It would be a big step to get the monied interest, the political interest, the corruption, the stranglehold of that from the scientific process, but I think that it is at least potentially possible. We can use the scientific approach, but a more holistic science. There is hope for that. That will give us some clues over the next century as to how to proceed.

In the meantime, I believe that we are not just looking at licensure but also things like standardization of medicinal herbs. What is a standardized herb? What is standardization? There is not a standard for standardization. The whole notion of that runs counter to having a living complete organism, a plant, that you can relate to as a whole organism. How do the two interface? That is a large part of the holistic aspect of herbal medicine. The intrinsic dialectic in medical science is this "let's reduce it down to one variable and control that variable," which is what turns herbs into drugs and turns healers into Western doctors or allopaths. I believe that the basic value of the American Herbalists Guild, at this time, is that it provides a venue or an arena where these types of discussions can take place and this type of inquiry. But, as far as licensing herbalists, I think that because of how things are structured politically right now, that probably would be the way to go. My only hope is that the educational standards that feed herbalists into licensure do not lose these really profound roots of their tradition.

Chapter 24

Deb Soule

Deb Soule is an herbalist, organic gardener, teacher, founder of the herbal apothecary Avena Botanicals, as well as the author of *The Roots of Healing: A Woman's Book of Herbs* (1995). She grew up in the western foothills of Maine where she first began gardening and learning to identify the wild plants growing in her backyard.

Deb founded the nonprofit educational center Avena Institute in West Rockport, Maine, which now has a large medicinal herb garden open to the public and sponsors various herbal classes. Deb tends an acre of medicinal herbs, sees clients privately, collaborates with several women health care providers throughout Maine and beyond, and is a student of biodynamic agriculture.

I spoke with Deb in her herb garden on June 19, 1999. I asked her about her experiences with endangered and at-risk medicinals.

OBSERVATIONS OF ENDANGERED PLANTS

I would say certain plants are endangered. I can say both from my experience of growing up in a rural area, as well as from wildcrafting herbs for over twenty years. What I have been watching is, first of all, of course, huge amounts of developments that have occurred recently. I grew up in western Maine. I have lived in the Rockport area for fifteen years. I used to spend a lot of time gathering.... I actually used to gather some valerian over in the South Paris area, and there is no valerian left *at all*. That is not even an endangered plant, but it is an example of the impact that development has on an ecosystem. I think this is far beyond our imagination. When you start to disrupt large tracts of land with big developments, you begin to disrupt more than just the plants there.

As I spoke earlier today [during the garden tour], the pollinators play a very crucial role to the survival of plants. If you lose the one, you have no plants because they need the pollinators to survive. I think that development and logging have had an impact on both.

What I have also seen is the practice of trading wetlands for development. In Maine, people can buy two wetlands and protect one wetland and build on the other. It is very strange. I mean, who should own a wetland, anyway? That creates a problem with, for example, calamus root. It is not a plant that I have ever seen in large amounts, but it is a plant that I have seen near rivers and wetlands that are fairly undeveloped. I have seen it less frequently in populated areas.

BIOREGIONALLY ENDANGERED PLANTS

You have to look at the endangered plant situation bioregion by bioregion. For example, goldenseal wasn't even listed as a plant that ever even grew in Maine, but there was a little bit in Vermont, and as you head south into the Appalachian country, you will be amazed at the stands of blue cohosh and wild ginger and American ginseng and goldenseal. What you have to take into consideration, as far as the issue of whether a plant is endangered or not, is whether it was in an area to begin with and also in what kind of abundance it was there. And then, what happened?

I have a book put out by the Jocelyn Botanical Society in Maine that lists every county, and it goes through various species of plants found in those counties. I also have lists from various plant conservation groups in Maine. They list who is endangered, who is less endangered . . . they probably call it "at risk" . . . and then who is plentiful. They will say don't pick and pick freely. That is the language they use in those publications. Ginseng used to be here in Maine, and it is not here in Maine anymore. Supposedly there is a little area in Bryant Pond, Maine, where maybe there are a few American ginseng plants. It was once here. Blue cohosh actually . . . I don't know in what prevalence it grew here in Maine, but it is definitely on the endangered plant list here in Maine, as is black cohosh.

Some people would say, "Pleurisy root is not endangered here." "Trillium is not endangered here." It has been a huge issue for us on the UpS board to decide who is really at risk here. What are we basing our numbers on? It has been important for us to make contact with plant conservation organizations and field botanists who have been doing some of that counting. Pleurisy is on the endangered plant list in Maine. It used to be here. I have never seen it in the wild. I have fifty plants in my garden. We are saving seed and giving it to people. It is not an invasive plant. It is not going to cause problems by putting it back in the wild. In some other bioregion, it is not endangered. An interesting issue that has come up for United Plant Savers is people saying, "This plant grows all over where I live." Well, it used to grow all over here and now it is on the endangered plant list here. We have to be looking at our bioregions and who used to grow here, who grows here well, who has the potential for being restored back into its natural environment. And not to make such blanket statements like all herbs are endangered. Not all pleurisy is endangered. Basically what UpS has done in creating its at-risk and to-watch lists is collect every state's conservation list. So they are going state by state, not by bioregion.

ROLE OF ALIEN SPECIES IN THE LOSS OF HABITAT

The other issue too, of course, is the moving in of so-called alien species of plants and how they push out native plants. An example of an alien species is St. John's wort, *Hypericum perforatum*. It would be interesting to go back and see if anybody could historically say how it got here, but we know that European settlers coming to the so-called New World, stitched seeds into the lining of their coats and carried them in their boots. They brought their seeds with them. St. John's wort has become such a rampant so-called weed problem out in the West . . . from the Rocky Mountains to the Pacific Northwest that they introduced a beetle to wipe out that plant. That is an example of a plant that is foreign and has naturalized itself, and somebody has decided it's a weed problem. So, now they have introduced a bug that is a natural predator to that plant, but that plant wasn't here in the beginning. So, we are introducing a foreign bug into an ecosystem to eradicate a foreign plant and creating problems that we don't even know yet. That is one example.[1]

Another example is coltsfoot. It is a major problem in Nova Scotia and New Brunswick. They now spray horrible herbicides along the roadsides to try to kill coltsfoot. It is invasive into streams, wet areas, sandy areas . . . a lot alongside the road.[2]

Then, of course, a seriously invasive plant is purple loosestrife. That is a really serious problem.

ROLE OF THE HERBAL INDUSTRY IN ENDANGERED PLANTS

I don't follow the industry a whole lot. I kind of keep an eye on some of the things that UpS has been publishing in [its] newsletters, and also what Steven Foster is writing. He is an old friend of mine. I kind of keep up with what he is writing about because he is keeping track of the tons of particular herbs that are being shipped overseas or to particular pharmaceutical companies in the United States. I think that somebody like him, and UpS, and World Wildlife Fund play a role in monitoring the herbal industry. The World Wildlife Fund now has a whole division for medicinal plants, TRAFFIC USA. So at least there are a few people watching numbers of things.

I do really think that the industry has played a major role in the extreme decrease in goldenseal populations and also black cohosh, which you have probably read about in the news. There [are] tons of black cohosh going overseas into Remifemin, a menopause formula, that is produced in Europe. Goldenseal and black cohosh are two plants at the top of the list right now as far as "endangered" due to commerce. It is really obvious that massive amounts of those plants are being used by the industry. We have done that with other plants, too, for example, *E. angustifolia*. American ginseng, of course, has become a problem because harvesting was and is a livelihood for people in the South and a lot of it—most of it—was shipped to China in large amounts. So, I would say that industry does play a role.

At this point, we are in a time where the bottom line for so many people is money. The relationship to the natural world and to the health of the ecosystem as a whole is not at the top of the list for most. The herbal industry has continued to grow at such an alarmingly fast rate, which is scary to me. Go and look at these herbal products. Even St. John's wort, it is added into juices now and things like that.[3] It is just like, please. That is where the faddish part of herbalism is dangerous as far as people having no relationship with the plants. They don't have a clue as to what these plants look like to begin with and no, or little, relationship to the fact that these are amazing living beings that have as much of a right to be here as we do. Most consumers, unless they have the opportunity to take the time to really look deeply and educate themselves, . . . don't really understand it. I look at all of those vitamins and minerals and capsules and things that have herbs in them. I wonder where are all of those herbs coming from? It is scary to me.

GARDENS AT AVENA BOTANICALS

One of the reasons I wanted to have this garden open to the public is because I wanted to begin educating a more diverse kind of population. Perhaps they would come to visit here and hear me talk or read something that we have written, and maybe it might plant a seed for them. I still think that we are pretty much in the minority as far as real hands-on touching and respect for the Earth, and respect for the natural processes that go on around us every moment.

OUR CULTURE'S RELATIONSHIP WITH EARTH

The germ theory had a major impact on how we view our world. I think it has created people's phobia around dirt and earth and bacteria and viruses and all that kind of stuff. I think it has had a major role in our relationship to living, pulsing things. "Let's just kill 'em." It wasn't until World War II that pesticides, herbicides, and insecticides really came into being, partly from the leftover residue from the war. Juliette de Bairacli Levy and Helen Nearing were here about eight years ago. They gave a public talk at my house. About eighty people came. Some people asked them their feelings about the future and they both felt despairing, actually. It was interesting, their response. It wasn't a cheery hopeful answer.

Most of the time I feel that during the daily activity of living, I do my best to look at the beauty right in front of me and then sometimes I think about the bigger picture and a longer view. Then, I feel like we have a lot of work ahead of us. It is huge, and it is discouraging to me, I have to say right now.

The genetically engineered thing is a part of this distance from nature as well. The monarch butterfly piece [Losey, Rayor, and Carter, 1999] just came out in *The New York Times*. There was some inside testing done on BT corn with the monarch butterflies and larvae. Over half of the monarchs died when they came into contact with the pollen of the genetically engineered corn. I see us moving in these directions, and it is terrifying. It is going to have an impact on the natural world that we can't even begin to see. It is going to affect our native plants; it is going to affect our naturalized plants; it is going to affect us. What kind of an effect is it going to have in a garden like this? It could be completely out of control and we don't even know yet? And we are introducing it into nature? That is where my hopelessness is. . . . I feel very hopeless as far as how to respond to that.

We can get discouraged easily when the weight feels overwhelming. For me, working in a community like this, there is an energy. When you are always working by yourself, you can tend to get discouraged more easily than if you are in a community of somewhat like-minded people, even though there is diversity here amongst all of us. It can help hold a more positive energy. I think that is part of what we are trying to do in Avena. I think that is why people come here. There is something that draws them here because we are trying to educate and offer a more holistic viewpoint. We have our work cut out for us.

DOMINO EFFECT WITH ENDANGERED PLANTS[4]

We can reduce a plant to just its chemical constituents, but perhaps what we need to be doing is looking at the bigger picture of plants that grow where we are and why is it that we use certain herbs and not others. For example, the slippery elm issue—people are trying to use marshmallow root instead of slippery elm. We are after demulcent properties or the berberine properties. . . . I think that it is important to keep that in perspective, but I also think that we have to think a little bit bigger about it. The difficulty, and maybe that's again where the paradigm shift has to occur, is in helping to educate people that it is not about using St. John's wort in place of Prozac. It is not about using gold thread instead of goldenseal because of the berberine content.[5] It has to be a bigger paradigm shift. I think we have to include more than just the constituents in our reasons why we choose certain herbs. I think it is important to understand that aspect, but I think we have to become a bit bigger. I don't know how you educate the general public about that.

HOLISTIC VERSUS ALLOPATHIC USE OF HERBS AND EDUCATION FOR THE GENERAL PUBLIC

One issue for me is shifting the paradigm away from the biomedical model or high-tech model of medicine. I think we have to really look at our use of plants from a much bigger perspective, instead of just replacing a drug with a plant. That is what is happening right now. It is rampant. Basically, all of your corner drugstores and large chains have herbal capsules in them. It is because people are asking for them.

How do we educate the general population? Really what we are talking about is how do we shift from a very allopathic, scientific way of looking at medicine and healing to much more of a holistic approach to living that includes everything in our daily lives from the way we get up in the morning to the way we go to bed at night, the food we put in our bodies, the love in our lives, the satisfaction in our working lives, and the environment we work in, even the cars we drive. To me, it is huge.

My work is with herbs, but there is a huge opportunity to shift the medical model in this country, not to exclude allopathic medicine but to create much more of an integrated system which includes everything about how we choose to live.

I think that trying to educate the general public about the basic use of herbs on a daily basis is really talking about how do you nourish yourself on a daily basis. That is the place at which I begin to teach about herbs. So many people are viewing herbs as a quick fix, and we have to get away from that way of thinking. There is so much more than that. To me, when we really begin to shift a paradigm, the possibility of effecting change in a lot of different areas could happen. As you shift one thing, other things shift too. I find that very exciting and feel like my life is dedicated to that.

I am already seeing some change. I think we really have to reeducate ourselves. We start with ourselves and our families and our communities, on very grassroots level. Certainly people like John Robbins, I respect him so much, but his books are radical compared to what a general person is going to read. We need his voice and we need all of the voices in between.[6]

It is not about cutting out one or another healing system because allopathic medicine has some very important things to share. I would include it as a part of a more holistic model of medicine—homeopathy, herbs, allopathy, body work—it goes on and on, the list of therapies that are helpful to us. Allopathic medicine has been the major model, and it has a very strong foothold. It is deeply rooted in the Western world, and it is very connected to the petrochemical industry and conventional agriculture. They are all very interwoven with each other. If you begin to pull the footing out from one of them, it will affect all of them.

VARIETY AND BALANCE IN THE HERBAL WORLD

We need every variety of herbalist in this movement from the clinician to the gardener, but I do get a little concerned with herbalism becoming too scientifically oriented. We need everybody figuring out how to talk with each other and work with each other and offer the best of our experiences to each other. This is where herbalism, to me, lives.... It's in the gardens and in the forest and the meadows and the fields. An herbalist sitting behind a desk at a computer all day long ... that's great for some of your work, but I think you miss out.

I also really think we need the female voice in the world of herbalism. I think certainly Rosemary [Gladstar] has helped to keep that alive. I have to really respect her because she has been a strong voice for twenty years. Certainly, in organizing herbal conferences, she makes sure that there are plenty of women teaching. I appreciate that. She tries to have all the worlds represented as best as possible.

I think we need to listen to the people who have lived amongst the herbs forever and ever. I am much more interested in what they have to say and their hands-on experiences. The extraordinariness has to stay alive in the midst of blending some of the scientific information that we are able to have in this time. I think we will all benefit from that.

HER WISH FOR HERBALISM

I would like to see herbalism stay in a grassroots place, and I am concerned about that. I am concerned that it is becoming a multinational, corporate thing.... I think there is a strong grassroots movement in this country, and I think that we should continue in this direction, keep that alive, and keep it spreading in a good way. I hope that is what happens. Herbalists have to be out there with our hands in the dirt talking to our neighbors and our communities. To me, that is where real change happens. It happens community by community. If you have enough connecting going on, you can change policy. I am here in this garden because I want to engage the public in conversation. Maybe they will go away thinking about something a little bit differently. Maybe they will decide to buy some organic meat next time. I recently had the opportunity to write a piece for a nonorganic gardening magazine. It was on why I choose to run my business using only organic herbs and why I choose to eat only organic foods. It is little things like that I hope have a little effect. It is not going to change all at once, but we begin step by step.

Chapter 25

Sharol Tilgner

Sharol Tilgner is a licensed naturopathic physician in Eugene, Oregon. She founded and, at the time of this interview, was the president of Wise Woman Herbals, Inc., an eleven-year-old company that supplies more than 300 herbal products to health care practitioners. Sharol was the director of the Portland Naturopathic Clinic pharmacy for two years. She also spent three years molding an old cattle ranch into an organic herb farm that is now called Shalimar Gardens. She is an associate editor of *Medical Herbalism* and has produced two herbal videos titled *Edible and Medicinal Herbs,* Volumes I (1991) and II (1993). Sharol lectures at various colleges and conferences across the United States. She also promotes and sponsors the Pacific NW Herbal Symposium each spring. Her most recent contribution to the herbal world is her book *Herbal Medicine from the Heart of the Earth* (1999). She is currently working on a new herbal video series, teaching classes, and developing a new herbal educational farm, Wise Acres.

Sharol and I spoke over the phone on September 29, 1999. I asked her about her experiences in the business of herbalism.

HER WORK WITH WISE WOMAN HERBALS (WWH)

Our mission at WWH is to supply quality. That is the basis of everything that we do. Anytime we are making a decision, that is always what we come back to: Are we sticking to what we are attempting to do here? I think that is probably the main thing about having some sort of a mission; it keeps you on track. As far as what I do, I have done just about everything there is possibly to do with the company. I started the company initially in my kitchen. At this point we have six thousand square feet here, and we are at a point where we have to grow again. We have to add on probably another fifteen hundred, two thousand square feet. During this period of time, I have done everything from wildcrafting to starting a farm. Actually, we sold the farm to someone who is continuing to supply us with herbs, but we started a farm from scratch. I have also been in charge of making the herbal products, advertising, sourcing things, buying equipment, managing employees—anything that there is possibly to do as a company. I may not do all of it at this moment, but it has happened at one time or another.

HOW WWH BEGAN

The company started by accident. It wasn't a company where somebody decided that they were going to start a company. It was more in the line of how somebody makes cookies in a kitchen, and everybody says these are really good. Then they give them to friends and family. Everyone says, "Oh you should be selling these or doing something with them." That is basically how the company started, except that it was more I had given so much away to my friends and family and they couldn't take anymore. I still had all of these extra products because I love to make herbal products and I love to wildcraft and grow. So, I started approaching people that I knew who could sell my products. That is how the business started.

At one point about a year after I started selling products, I thought, *Well, I kind of have a business, I guess I should have a name, and maybe I should actually have a label.* The business wasn't created on purpose so it was never a "I have to make a profit" type of a situation. It was always done on "This is really fun, and I like what I am doing." That is how I ended up with four hundred fifty products. Somebody would ask for something, and I would think, *Oh it would be fun to make that. I have never made that before. I could learn something.* The next thing I knew I had all of these products.

THE AUDIENCE FOR WWH

We sell mostly to physicians. We have some pharmacies and some health food stores, but the business started out of the naturopathic community. That was the community that I was in. I was actually going to naturopathic school when I started. It started there, and I would say that eighty-five percent of our customers, maybe eighty percent, at this point, are naturopaths. Then we also have holistic medical doctors, chiropractors, acupuncturists, herbalists, then the occasional health food store and pharmacy.

The health food stores have a whole different bent about what they are interested in. Our interest is always in quality; their interest is largely in what can they sell, does it have a gimmick, what does the label look like. They have more of an interest in that than they do in what is in the bottle, which is surprising, but that is what I have found. They want [a product that is] cheap and looks good and has something entertaining to do with it. It is an entirely different market than we normally serve. So, when we started looking at that market and interacting with them, it was just too weird for us. Most of the health food stores we are in are those that have someone who really cares about quality.

CURRENT HERBAL "BOOM"

What we have seen happen is that with the rise in the market interest, there was a rise with our business also. At the same time, to some degree, there came a point about a year ago, where it seemed like things leveled off and dipped down even. I talked to other people who had herbal companies, and they were all experiencing the same thing. They said that their sales had gone down, and they were surprised. What I think it is that so many people came into the market because they saw it as a lucrative market that sales were dispersed.

I think that there are a lot more people using herbs and the interest in herbs is up. Some of the companies like Centrum are advertising on TV and going into Wal-Mart and places like that . . . my guess is that they are probably grabbing a bunch of people who are new to the market also. They are advertising to them and the people are seeing that and going to buy their products. Our market is more the physicians. I don't think we saw as much of a change in our market as other people that I have talked to. Some people have had huge changes. Their company was growing eight times as fast as usual.

There was less of a noticeable increase with our sales to physicians, although our retail interest did increase. But it is hard telling what that relates to. We also had people writing about us in different newsletters and magazines. That could have been it also. Without having a sales and marketing department to trace all of those things, sometimes it is a little bit hard to tell. We have just started paying attention to that information.

I think it might be hard to start as a small company now in the middle of this herbal boom. I think that it depends on what you want to do, but it would be much harder because there are so many people doing it now. When we started twelve years ago, there was such a huge difference at the time. There were other companies, but there weren't very many. Not like there is now. There is a glut of herbal companies now.

SOURCES FOR HERBS AND ENDANGERED MEDICINAL PLANTS

We use organically grown and wildcrafted herbs as well as biodynamic herbs. In fact, our first choice is usually to get biodynamically grown herbs and then organic and then wildcrafted. But it depends on the herb. If there are things that can be wildcrafted that there is a huge abundance of, that will be my first choice. If it is an herb that isn't in large abundance, then we go organic.

As things have become endangered we have stopped selling them. Sundew we quit selling almost right after I started the company. We stopped selling pygeum [*Pygeum africanum,* commonly used in men's health formulas], I remember, bryonia [*Bryonia alba*]. Hydrastis [*Hydrastis canadensis,* goldenseal])—we quit selling the wildcrafted three or four years ago and went to only organic. Things happen as we go along. We just watch and see what is going on and make moves either that we discontinue something if we have to or we try to find it organic.

Black cohosh—I have been telling everybody to grow it. We want it. That is what everybody else is doing too. So what happens is that people do that. Right now there is a huge amount of goldenseal being grown. There is becoming enough of it on the market now that the price is starting to go down again. We went from paying forty dollars per pound to eighty dollars per pound just like that when we switched from wildcrafted to organic. Now that it is down, I think we can get it for about seventy dollars per pound. Next year, I expect that it will be about sixty-five dollars per pound. That is what it is looking like at this point.

With our goldenseal products, when we transferred to organic, nobody else was, so we were competing with people who had prices that were way less expensive. We did have to increase the price, but there was no way that we could increase it to take care of the cost. We had to consider it something that we wouldn't . . . I don't know if we made any money on it or if we made very little money on it. I don't take care of that part of it anymore. The person that takes care of how much money we are actually making on it was looking at it and saying, "This is not good." At this point, it is getting a bit easier because the price for the raw material is going down and other people are starting to switch to organic now also, so you see their prices are starting to go up. It makes it a lot easier that way. But we did have people who didn't want to pay the higher price. We did lose some people.

My biggest concern is that we are going to keep seeing the plants disappearing. It is the same thing with all of our resources that are disappearing. We can keep being really careful about how much we use. We can conserve. We can make sure that we are planting things. And with other types of things beyond plants, we can make sure that we recycle. But it keeps coming back to the same thing. We have a huge population on this planet. We are at six billion now.

IDENTIFYING PLANTS FOR USE IN MEDICINES

I think that American Herbal Products Association [AHPA] has been working for some time on coming up with some good manufacturing practices [GMPs].[1] One of the problems that has come up has been that there aren't tests for everything. For instance, you can't just bring everything in and do a TLC [thin line chromatography] on it or gas chromatography. You can't test everything that way because there are not standards to follow with every plant. Then it is also very expensive to do that.

Why do tests if you know what's . . . it is like bringing in a carrot. I know that is a carrot. It looks like a carrot; it tastes like a carrot. I can tell it is good quality or [if] it is really poor, it is limp and old. Everyone knows what a carrot is and many herbalists can judge quality when an herb comes in. Yes, they know it is the herb it is supposed to be, just like you would know a carrot.

So, there is a subset of AHPA—well, it is just a little group—that is looking at standards of the industry for small companies. They are looking at organaleptic techniques for identifying quality, which basically means looking at it, smelling it, tasting it, feeling it.

Ultimately, I hope there will be different accepted ways to identify things. Let's say that you get a plant in and you are checking it out. You would have a sheet and on this sheet you would have different methods you could use to test it. One of them might be microscopic, so you take a section of it, look under the microscope, and it has all of the characteristics that it is supposed to have. Then you look at the whole plant and it looks like what it is supposed to. Then you smell it; you taste it. You are pretty assured at that point that you got what you are supposed to, but if you wanted to, you could even go on and do gas chromatography. There are some plants that you might get that would be hard to identify that way. Maybe it wouldn't be as easy so you would have to do some other type of test. Or you might want further proof for somebody who is buying it.

I think if you use microscopy with it, though, that it becomes a little bit more realistic to people because it is a scientific technique. That is basically people's religion, it seems nowadays. If you add that to the standards then it might be more acceptable. That is what I believe the AHPA committee is doing right now. They are looking at that too.

FEDERAL REGULATIONS AND SMALL BUSINESSES

The FDA has not yet come up with the GMPs. They have been working with AHPA on that. What we do at WWH is we follow basic restaurant sanitation guidelines. At one time there was talk about creating pharmaceutical manufacturing guidelines for small herbal business people, but they're really too stringent for our industry. It would cause a lot of problems in our industry to do that.

One aspect of regulation that has affected us is labeling. The Dietary Supplement Health and Education Act has made our labeling much more complicated.[2] As a dietary supplement we all have to change our labeling to incorporate thirteen items on our label showing how much calcium, how much fat, how many calories are in our products. Most of these in herbal products at the dosages recommended come out as zero. Occasionally you have something that actually does have something in it, but they are mostly zero. It has been an incredible amount of money that the industry has had to spend on changing their labeling, looking at their products, and figuring out how they are going to do this. We had to reorganize our entire database. It was just crazy for us to have to do all these things. And our customers don't even care. They couldn't care less.

Now we have this whole system in place where we can fax or mail product information or for some things they are actually on the label. We had to change our labels for other things that are really tiny, like one-ounce bottles. We couldn't put it on the bottle so we had to put an eight hundred number on there so people could call. Now, we have a whole computer system set up for us to be able to fax and mail this information out to people.

There was a clause in DSHEA that if you were under a certain number of employees and a certain amount of money that you didn't have to do this. When we first looked at it a couple of years ago we were right on the line. We decided to go ahead and do it because we are going to have to if we got any bigger. I don't know if they are going to keep that clause in there or if they are going to phase that out eventually.

ROLES OF SCIENCE AND FOLKLORE

Well, we have lots of research here. I have file drawers full of it. I actually love reading research, although it is always secondary to me to the herbal wisdom that gets handed down through the ages. A lot of the research basically just says the same thing that people already know, but it is presented in a different way. I have always found it fascinating to read the information. I just got through putting a book out. It will be back in a couple of weeks from the printer. In the book, I have a lot of research. Basically what I did is I wrote the book, and then I went back and I put the research in. I knew people would want it. For myself, the research didn't matter at all. It wasn't even important, but I realized that people really would want to see that, to see that there was something standing behind what I had said. That was the way I approached it in there. I knew that there would be many people who otherwise would feel like there wasn't anything to support what had been said. They would not believe it. Same thing with customers who need research. We have plenty of research here and we can supply that to them, as long as there is something on it. There are some plants that have no research on them at all.

If you want to get right down to the heart of the situation, for me, I feel like there are a lot of people who are very disconnected with the Earth. You see it in how we pollute the air and the soil and how we are overpopulating the planet. I feel like the interaction with herbs and relying on the scientific data is just another symptom of that whole situation. I really feel like people need to reconnect with the Earth and feel more a part of the Earth and realize that they live off the Earth. They are a part of the Earth's body, and they consume things off of it. Learning about herbs is one way for them to realize that. They can go out on herb walks, and they can connect with the plants and realize that they are incredible little beings that harness the sun and give us this energy. By doing that they are making a reconnection. I feel like it is really important at this point on the planet that people do that. So, I see folklore as having not just a huge place within herbalism itself, but I see it as having a huge place in saving ourselves, I guess, in a sense.

How can the dominance of science change herbalism? Well, what I see happening is there are people who use herbs in a very allopathic way in that they are using an herb to treat a disease process. In herbalism traditionally, herbs are used to treat a person. You know an herb really well, and you understand a lot of different things about that particular herb. Then you have a person with a lot of different things about them that you find out. You see a whole picture of that person. Then you fit the herb with the person. What is happening now with a lot of the scientific research is it is looking at taking care of disease processes. Herbs are being fit into very much drug type categories, in that if a person has this particular illness then you use this particular herb for it. I think that is a real big shift in that kind of a way.

I think that this is changing and maybe it is just my minuscule world that I see it in. It seems that people are feeling very separate and disconnected, disjointed and in quite a bit of pain. They are trying to bring meaning to their life. You see a lot of people who are grasping for this. I feel they do want something different and they do want a different way to relate to the world. When you talk to people about herbs in a more holistic type of a way, an earth-centered type way, many people really love it. People you wouldn't even expect that would love it become fascinated by it.

BRINGING HOLISTIC CONCEPTS TO THE MAINSTREAM

I think that people have to feel comfortable first of all in what they are doing. They need to feel that they are not going to be stepped on in any way or told that they are stupid or don't know what they are doing. Then once they feel comfortable with where they are at, we can help them to open up and see something else, in the holistic concept, and include something else with what they are doing. It started in physics when researchers realized that they were not separate from what they were observing. They brought that to everyone's attention. And told people, "You know, we can change how the atom is moving" [laughs]. I don't even know if the whole world realizes that took place many years ago, but quite a few people do. That set off all sorts of books that came out. The *Tao of Physics* [Capra, 1975] and all sorts of different things that appeared that really caused a shift in people's consciousness.

Chapter 26

Peeka Trenkle

Peeka Trenkle is a clinical herbalist and natural health counselor specializing in women's health, fertility, pregnancy, and children's health. She is a certified childbirth educator and a professional member of the American Herbalists Guild. Peeka is on the faculty of the New York Open Center in New York City where she teaches Green Medicine: An Eight Month Training in Western Herbalism and The Parenting Series: A Holistic Approach to Raising Children. She is the coproducer of The Art of Birthing, a conference exploring the philosophical, sociocultural, medical, and spiritual dimensions of the birthing experience.

Peeka and I spoke over the phone on October 8, 1999.

HOLISTIC AND ALLOPATHIC HERBALISM

We have two things going on in our culture right now. We have allopathy, and we have the uptake of herbal medicine into the mainstream, which is being called holistic but it isn't holistic. It isn't taking into account the health of the whole person within the context of their environment. So, holism is really the health of the whole . . . the dynamic health of the whole person within the context of [his or her] environment. That belief, that construct, is part of what is called the vitalist tradition.

What I see as being a real problem today in herbal medicine is that there are many very skilled, very brilliant herbalists who are positioning themselves to work in a very medical way with doctors. They do this because herbs work well in that system to heal diseases. But it is still looking at that disease as something that comes from the outside.

There are very basic issues up for us that need to be discussed—mainly that we need to look at what is disease, and we need to look at what we are doing as herbalists. The way that I work is in trying to support the vital force of the person, to support the health of the person so that the body resolves things through its own abilities. It is a dynamic thing. It is always a dynamic, moving thing. I really think more as a homoeopath than an herbalist. I think that the whole way of viewing health has been kind of forsaken by the herbalists.

Holism is the idea that the body heals itself, that what needs to happen is either the cause for disease be removed or the body be supported in some way. Sometimes herbs need to be used to actually heal tissues, but ultimately it is the body that does the work. The herbs facilitate that work. They are a catalyst or a nourishing agent or an energetic . . . an energy supply, if you will.

What is being called holistic is just using herbs the way drugs are being used. It is using the same medical construct and instead of using nystatin for yeast infections they are using calendula or thuja or monarda. It is still a certain classification of herbs for a certain classification of disease. It is very reductionist.

People who use herbs in this way and call it holistic don't understand what they are saying. What they really mean to be saying is natural remedies. There is nothing wrong with that. It is certainly a gentle, safer approach. I don't have a problem with that, but I don't think they realize . . . I think holistic has been bandied about and has come to mean, commonly, just noninvasive, palliative remedies.

Allopathy is not what we want, but we are practicing herbalism as if it were allopathy. Botanical medicine is allopathic medicine. What do you do for a fever? You bring the fever down. You give a cooling herb. Well, in some herbal traditions, you give a warming herb and that is called a diaphoretic. That is more homeopathy. But allopathy is absolutely rank in herbalism. It is because that is how people grew up. One of the reasons I like to teach about vaccinations is I think it shakes people's minds up, and it gets them thinking about this whole germ theory, and the way that we have been taught to view illness, and what we believe needs to happen. But there are so many people who are at that elementary level. "I have a cold. I have to stop the runny nose and maybe boost my immune system." That is about as far as they go. It is very tiresome. I get frustrated just because I see this . . . almost herbs by numbers. The magazines are all doing it. Black cohosh for menopause. I haven't known one woman for whom black cohosh really worked for menopause. It is usually more of an adrenal thing.

HERBAL EDUCATION

Herbal education is hodgepodge. Now, I happen to believe that is a good thing. I happen to believe that herbalism is of the people, by the people, for the people, however the people want to do it. I don't think that herbal education needs to be standardized the way it is in homeopathy or in medical school. I don't think there need to be these systematic schools set up to teach the body systems unless you are drawn to that route. They can be there, but it shouldn't have to be for everybody. Herbal medicine is like cooking. You can go to school to become a chef or you can make bean soup at home and bake your own bread. You are still using food, and you are still doing nourishing things. It is just the level of expertise. Not that one is better than the other. Other systems cannot maintain that flexibility, I mean chiropractic or homeopathic or acupuncture or any of that. You need to have systematic study. There is a system in place with those modalities. There are laws of healing that have to be followed. But herbalism is different. It is much more how a person lives. I think it really does need to be equated more with lifestyle. I think it is a very free and open system and should remain that way.

I think that there is a definite desire on the part of the pharmaceutical companies to have most extracts standardized and have herbalists licensed or discredited. I understand the importance that people place on the study of anatomy and physiology. I would have to agree that if you are going to work with people in this day and age, you had better know your anatomy and physiology. People are going to come in and say I have this or that going on. They will have been given a name and diagnosis and you have to have a good understanding of what that is. That is a good thing.

Herbalists should have a very profound understanding of materia medica and sometimes that includes some understanding of biochemistry. Do they need to have a degree in biochemistry? I don't believe so. Plant chemistry, botany . . . absolutely. In order to understand the materia medica you have to understand how the plant grows, where the plant grows, what it does in nature, what it does in the human body. But again, I think that homeopathy can give a better model than can Western medicine. Homeopathy has a very rigorous training. You have to really understand not only the materia medica of plants, but of minerals and of the imponderables, the gases and the animal parts and venoms and all of those things.

I personally think that following the model of medical school for herbal school is a big mistake, but I do understand the attraction. I am sort of a both-and person. I think that both can exist. There can be schools the way there are naturopathic schools. For example, National School of Phytotherapy is very medical-oriented, and I hear very good things about that school.[1]

I think that there can be different paths for the herbalist. But yes, you could be a medical herbalist and work within the medical system. Yes, you could also be a granny midwife using herbs. You should not be discredited because of that. The same issues are coming up in a lot of different fields, particularly midwifery. Midwifery has such a big issue going on that women who are so well and highly qualified are seen as illegal and are persecuted and prosecuted. That is a very great potential for herbalists, if there is licensure for some and not others. That needs to be addressed before the whole process goes into being.

SPIRITUALITY IN HERBAL HEALING

Let's put it this way: I believe that everyone has a spirit and that spirit needs to nourished. It is up to that person and [his or her] journey and . . . karma what that spirituality will manifest as, but I think that it does play a deep role in healing. How can you remove depression if a person doesn't have a reason to live? How can you help [people] to achieve their health goals if they don't have any vitality in their [lives]? No reason to do that. The whole reason to get healthy is that you have a contribution to make in the world, and spirituality always has a part in that. But everybody's spirit is different. I have a relative, for instance, who is probably one of the most deeply spiritual people I know and lives all the principles of, say, Buddhism. She is generous, and she develops herself, and she is kind. But she calls herself an atheist. It is not always about having a personal awareness that you are cultivating spirituality but that your spirit is intact.

People like Andrew Weil or Christiane Northrup have been great bridge-building people. They really have helped the mainstream start to understand the necessity of caring for the spirit more, but it gets so trivialized. It gets so trivialized because basically the culture wants things to remain status quo, and, frankly, once people become healthy they start to look around, and they say wait a minute the world is really screwed up. It is a conflict of interests for the medical industry to create truly healthy and autonomous people and to facilitate spirituality in people.

Americans are so divorced from reality and really seduced by this kind of consumer-driven mentality. You even see that in the herbal community. Not so much at Green Nations.[2] That is a very different situation, but think of any health food store you walk in and, yeah, we have got more herbs there, for sure. However, it is all packaged. It is all in bottles; it is all in plastic containers. It is so much garbage. It is so much consumerism. It is so expensive. That isn't herbalism. Herbalism is what 7song does. He lives on the land, lives with the Earth.[3] Herbalism is living your life in the simple ways of back to nature. That is how I see it.

HERBALISM AS CONSUMERISM

How could consumerism be damaging to herbalism? Well, just purely ecologically. How much land is getting torn up for herbs now? How much packaging is getting dumped? Those two issues alone are hugely devastating. See, when I became interested in herbs, yeah, you had to go buy them in a health food store, but the whole point was that you read everything that you could possibly find. You were doing this on your own. You were taking responsibility, and you were learning something. You didn't go to an expert unless you were very lucky to find somebody who knew more than you did. You didn't go to an expert to learn about herbs. You just sort of figured it out on your own. You kept going with that. Now people feel sort of beholden to the health food store owners. They feel beholden to the herbal teachers. A lot of that gets diffused by the quality of teaching that is going on. There are so many good teachers out there.

I am not saying that all bad things are happening. There is a lot of good that is happening too. But this idea that well, what is good for this? If I have arthritis, I am going to go to the health food store and I am going to ask what herbs should I take? The salesperson is going to look it up in a book, and he or she is going to tell me. How different is that from the way people have a pain, and they go to the doctor and say what can I take? Whereas holistic treatment of arthritis would ask, why am I hurting? What is going on in my life? What am I doing? What have I done? What is my history? Who am I? Why my leg? Why not shoulder? Thinking it through. Unpeeling that onion. Finding a way to regain that dynamic life force that makes you stable.

Consumerism takes away. It does two things. First of all, there is that whole thing of you can't have freedom if you don't take responsibility. That is a basic premise. You don't get the responsibility or the freedom by going to a health food store and buying your herbs. You do this if you already embody the responsibility. I can go to the store and buy ginseng if I need it because I am flying somewhere. I can use that. It is nice that it is conveniently there, but I can also do without any products. I can also go in my backyard and make myself something from what is growing there. My vision of herbalism is that we have more food shelves and herbal endeavors, group endeavors and have people with their own herbs in their backyard. Those people who live in the cities can all share a community spot and grow their own and make their own. Or they have their neighbors make their medicines. We start to manifest more community with each other and work bioregionally to make that part of our lifestyle again, to make the Earth part of our lifestyle again.

HOW DO WE GET TO HER IDEAL HERBAL VISION?

The first good step is really the kind of herbal education that is going on. I think that the teachers in this country are really committed. If we don't hold the same vision, at least we support the greater view together. When we come to the big issues of ecology and the future of our world, we pretty much see eye to eye. That is number one. Education is number one.

And then, there is making a commitment. I think about when I teach in New York City. I always will bring into my classes resources about the herbal farms that are in our area. I will encourage people to go and visit those places, and I will encourage people to start getting their herbs only from these places so that they are not buying herbs that come via Massachusetts from New Mexico, or buy things that come through via Colorado from India. It is like, why? When we have everything that we need in our bioregion.

I had a discussion with 7song. We had this conversation about how I have made a commitment to go to homeopathy school. I have probably done this mainly because that is where my sensibilities are very strong, but also because ecologically we cannot go on using so many plants. In homeopathy, you can pick one plant and make medicine for a million people. He said, oh, that is terrible. You mean you would only have to pick one plant? And actually, I support his vision in the long term—that everyone should live on the land, everyone should have access to the plants, should know the plants, and be able to use them for food and nourishment the way human beings are meant to live. In the interim, however, we have a lot of healing to do. People are really screwed up in their thinking. There is a lot of avarice, a lot of greed, and a lot of disregard and disrespect for people and the environment.

LICENSURE AND LEGITIMIZATION

I would have to say, I really question the licensing of herbalists. I think that a lot of discussion and clarification would need to happen before I would feel comfortable with it. I think that there possibly could be licensure for a certain kind of herbalist, particularly a medical herbalist. I could see that might be of importance, just given our litigious society. When you put yourself out there as a medical herbalist you are part of that gang. As far as the rest of the herbalists, pretty much ninety-five percent of the herbalism that goes on the United States is not medical herbalism, per se. For that ninety-five percent, I think no. It is a very bad idea because herbalism is . . . a self-care system. There shouldn't be licensure for people who are facilitators for self-care. Also, the fact is that there is no pathway to licensure in this country that has any understanding of vital health. So, how can you expect a system that is so corrupt that at its core are the health and human resource department and the CDC and the FDA and these people who have no understanding of even physical health. . . . I mean, for heaven's sake . . . there is no gestalt there. There is no understanding. There is no way that an herbalist would be understood. The kind of documentation, the kind of credentials and such that an herbalist would need to have would take them away from their work. There are so many people who, just because of how they live and what they know, are able to help people enormously. Myself, I am not licensed. I am completely classified as a layperson. But I have a very thriving clinical practice. I even hesitate to call it a clinical practice anymore. It is a consulting practice. People come in to learn. I have people coming in for all kinds of things. They learn, and they take that information and apply it in their lives and, lo and behold, lots of things happen—good things. But it is a system where it is the individual doing the work. It is a very different system. I just think too that there would be herbalists excluded and possibly punished for being illegal. There will be those who are licensed and those who are not licensed.

You go to conferences, and it is not so much the herbalists themselves but all the people that come to learn who are grasping for information . . . let me learn this. What is this? What is this protocol? Rattling off acts and who knows more than whom. That is awful. I have seen this with my students. The only class that I got a bad evaluation in was my advanced class where I was teaching case taking. I kept it at this very human level. I didn't go into the protocols for infertility. I had some students in that class who very much wanted that information to fill up their notebooks. That is a track that is very attractive to a certain kind of mind. So, ultimately as a purist, I would say no licensing for herbalists, but I understand that there are differences in how people want to practice . . . and this is what I mean by practicing medical herbalism. It is really like being a doctor. It is not that you can't be extremely good at that and effect really good changes in people, but it still bypasses that premise that the body heals itself.

POTENTIAL FOR RIFTS IN THE AMERICAN HERBAL COMMUNITY

I do think that if we don't start looking at our common ground and include more and more people in that common ground, then we really do run the risk of getting wiped out. United we stand. It is really true. Another reason that I have decided to go to homeopathy school, and this is a very intensive program that I have taken on, is because no one is speaking to that issue. No one is speaking to the issue of, hey, how do herbs and homeopathy work together? How do they oppose each other? Where do they not work together? One of the big disasters of the nineteenth century is that the homeopaths and the herbalists never found any common ground. They were opposed to each other. I think that is certainly an important thing. But you will find that most herbalists, firstly, they don't understand homeopathy, and secondly, they really . . . if they do understand it a little, they dismiss it. The same way that doctors do. I think there is a lot of in fighting among all natural practitioners. There are herbalists who think that naturopaths are too medical, and naturopaths who think that herbalists are too amateur, and herbalists who think that chiropractors are just out there to make money. It is on and on. There need to be alliances. I think that the issue of licensure has the potential to bring all of these discussions out. That needs to happen. I welcome it even though I find it horribly uncomfortable. I really don't like sitting in a room of people who disagree with each other, but on the other hand, it is the only way.

MOVING INTO THE MAINSTREAM

You see issues in anything that used to be completely fringe. Herbal medicine used to be completely fringe. Home birth, forget it. Completely fringe. And homeschooling. I homeschool my kids, but in the last five years, but especially in the last three years, suddenly homeschooling is becoming hip. I am interviewed by magazines and homeschool newspapers. People are joining our group like crazy. But what happens then is that people come into the fringe paradigm, and they bring mainstream culture into the alternative and it doesn't work. We would have people coming into our homeschooling group telling us how completely disorganized we were. Well, for us, it worked great not to have officers and not to have one person in charge of everything. We just rotate the jobs and stay loose. But nobody could get that. The growing pains were horrible. We lost members over it. The thing is that it is just the way people think. There are so many people in the herbal community now who really didn't come into the herbal community because they were living an alternative life. They came into the herbal community because this is an up-and-coming market. This is a good way to make a living. I have students now who say, "I come to your class because I want to become an herbalist." It never occurred to me to say that I was going to be an herbalist. I just sort of did alternative, crazy, weird things. And herbs were part of that.

Epilogue

As winter fades to spring, and the leafless coltsfoot flowers begin to show, I sit here, my back against a great sugar maple in my yard. I sigh a deep, cathartic sigh: It is done. I have said what I had to say, and now I can move on. Certainly I have said a great deal already. Certainly the reader can hear my voice, my bias, my desire to live closely and in harmony with the Earth, throughout the preceding pages. I have woven together my thoughts and those of the herbalists whom I interviewed. But what one message do I want the reader to take with him or her when he or she finishes reading? If only one small piece, one sentence, can be remembered, what should it be?

Without getting too esoteric, I would like the reader to remember the concept of *beyond*. By that I mean that which is not part of our current thinking, but resides in the land of infinite possibilities. I use the term *beyond* because much of it seems inconceivable by modern, linear thought, but just around the corner if one switches one's perception ever so slightly. It is an idea, or rather set of ideas, which is beyond current imagination.

Some of the herbalists I interviewed spoke about beyond when they considered the possible merging of Western science and American herbalism. They noted that the two might produce a synergistic response that our imaginations cannot grasp at this time. The herbal community needs to search for beyond when considering possible licensing scenarios so that one might be arrived at which is equitable to the whole community. The possible effects of American herbalism's integration with mainstream culture is certainly beyond. It is beyond to think that American herbalism, especially the holistic paradigm, could play a leading role in grounding the mainstream culture, but it is just around the corner.

It is no secret that the mainstream market culture is drawing the herbal community inward. Like it or not, demand for natural and herbal medicine in particular is exploding. Mainstream America yearns for the knowledge and the wisdom of the herbalists. They desire connection to the natural world and some semblance of peace and stability in our fastpaced world. They need the wholeness that our fragmented world has stolen from them.

American herbalism centers itself on holism. Holistic herbalism (and to some degree botanical medicine), at least in theory, understands the importance of home and place. Using the tenets of the holistic approach to healing the individual, American herbalism may begin to heal our society, to defragment it. It may exalt the importance of tradition and legacy; it may temper the isolationist logic of biomedical science while utilizing the more positive aspects of it; it may connect people to their local ecology and allow them the realization that they belong to a place, that they are, in essence, growing from the Earth. The potential for healing through American herbalism is enormous. The "in" is already there; the people are just waiting for more, a deeper look into this field. Now is the time for those of us in the herbal community to examine our belief systems and find ways in which we can be truly integrative while maintaining the ability to shift paradigms in the mainstream. Now is the time for the herbal community to consciously direct its future.

Through careful examination of the issues within the American herbal community, this healing potential, this paradigm shifting, may be manifested. Through provocative, innovative thought and acknowledgement of the ideas of *conscious* and *unconscious power,* American herbalism may blossom into the tall, sturdy tree that it is meant to be.

As I lean against this towering maple tree, I envision myself as the tree. I sink my roots deep into the Earth; I put down the foundation that holds the weight of the reaching branches in place. I extend my arms skyward and stretch my shoulders as far as they can go. I feel the tips of my fingers sprout tender, new branches with delicate buds, which will open and later provide shade from the heat of the summer. I tune into the flow of sap from the earth to the sky: the sap flows through me; it nourishes me.

In this way, anchored by my roots in the soil, I begin to understand what it is to connect Earth and sky. But I cannot be one individual on my own crusade. I know the truth that anything which is to manifest meaningful and lasting change and healing does so with the help and support of an entire community, the action and energy of the many seeking a common goal. With that, my vision changes.

Now I am but one person in a pile. We are climbing around and on top of one another and finding comfort in wrapping our bodies around one another. We are forming the tree. Our roots again sink deep into the Earth, tapping into a sphere of nourishment, of life energy, at the core of the Earth; our branches touch the blue, cloudless sky. We are the community; we are the tree. We struggle and strive and reach, reach, connecting once again the Earth and the sky.

I realize, in this vision, how the Earth and sky are the beyond. Through our acrobatics, we reach into the beyond, the wealth of unimaginable ideas, and bring them into the now. The sap is the blood, the liquid of the beyond, and the tree, the community, gains life-sustaining energy from the drink. As we drink it, we become the guardians of the beyond. As we grow stronger, as we wrestle with issues of sustainability, we bring the beyond, the unknowable, into the physical world of the now. We bring it and show it to the masses that come to find shade from the burning rays of the sun. We share of ourselves generously so that others may also know what lies beyond, so that they too may stretch the bounds of their imagination realizing that beyond does not need to stay beyond.

That is my dream for American herbalism, for it to be the towering maple, for it to realize its enormous healing potential by harnessing its conscious and unconscious power. It is my dream that the American herbal community can work through the issues that it faces in a collective and peaceful way with concern for the ecology and community at the forefront of consideration. I dream that American herbalism may continue to develop its relevance in the modern world, reaching into the realm of the unknowable possibilities and demonstrating that those ideas are not as unknowable as one might think. My dream is for sustainability, unity, and respected diversity within American herbalism. With careful concentration and innovation, that dream may become a reality.

With the first, tentative drops of rain falling from the now cloudy sky, my reverie, my vision, is broken, and I leave it willingly, knowing that each time I lean against this tree, the vision becomes more real. Each time, I take a bigger and bigger piece away with me, turning thought into action in my own life. The rain becomes a stronger force now, rushing me inside the house.

As I sit inside and stare at the maple tree through the window, a warmth grows within in me. I have done good work. I have said what

I needed to say. I hand this over now as a gift to the community. May we be the tree. May we accept the challenge. May we drink heartily of the life-giving, life-connecting sap. May we sink roots deeper and reach branches taller until we fully realize that we are the beyond brought into the now.

Appendix A

United Plant Savers "At-Risk" and "To-Watch" Plants

The following lists are published by United Plant Savers, a grassroots conservation organization dedicated to the preservation of native North American medicinal plants. The plants listed represent "herbs which are broadly used in commerce which, due to overharvest, loss of habitat, or by their innate rareness or sensitivity are either at risk or significantly declined in numbers within their current range" (Gladstar and Hirsch, 2000, p. 26). United Plant Savers welcomes feedback on the lists. Send comments to UpS "At Risk," c/o Horizon Herbs, P.O. Box 69, Williams, Oregon 97544, or visit their Web site <www.horizonherbs.com>.

At-Risk Plants

American ginseng *(Panax quinquefolius)*
black cohosh *(Actaea racemosa; Cimicifuga racemosa)*
bloodroot *(Sanguinaria canadensis)*
blue cohosh *(Caulophyllum thalictroides)*
echinacea *(Echinacea* spp.)
eyebright *(Euphrasia* spp.)
goldenseal *(Hydrastis canadensis)*
helonias root *(Chamaelirium luteum)*
lady's slipper orchid *(Cypripedium* spp.)
lomatium *(Lomatium dissectum)*
osha *(Ligusticum porteri, L.* spp.)
peyote *(Lophophora williamsii)*
slippery elm *(Ulmus rubra)*
sundew *(Drosera* spp.)
trillium, beth root *(Trillium* spp.)
true unicorn *(Aletris farinosa)*
venus fly trap *(Dionaea muscipula)*
Virginia snakeroot *(Aristolochia serpentaria)*
wild yam *(Dioscorea villosa, D.* spp.)

To-Watch Plants

arnica *(Arnica sp.)*
butterfly weed *(Asclepias tuberosa)*
cascara sagrada *(Rhamus purshimia)*
chaparro *(Casatela emoryi)*
elephant tree *(Bursera microphylla)*
gentian (*Gentian* spp.)
goldthread (*Coptis* spp.)
kava kava *(Piper methystricum)* (Hawaii only)
lobelia (*Lobelia* spp.)
maidenhair fern *(Adiantum pendatum)*
mayapple *(Podophyllum peltatum)*
Oregon grape (*Mahonia* spp.)
partridge berry *(Mitchella repens)*
pink root *(Spigelia marilaandica)*
pipsissewa *(Chimaphila umbellata)*
spikenard *(Aralia racemosa, A. californica)*
stone root *(Collinsonia canadensis)*
stream orchid *(Epipactis gigantea)*
turkey corn *(Dicentra canadensis)*
white sage *(Salvia apiana)*
wild indigo *(Baptisia tinctoria)*
yerba mansa *(Anemopsis californica)*
yerba santa *(Eriodictyon californica)*

Appendix B

Resources for Further Study

Organizations

American Botanical Council
6200 Manor Road
Austin, Texas 78723
(512) 926-4900
www.herbalgram.org

American Herb Association
P.O. Box 1673
Nevada City, California 95959
(530) 265-9552
www.ahaherb.com

American Herbal Products Association
8484 Georgia Avenue, Suite 370
Silver Spring, Maryland 20910
(301) 588-1171
www.ahpa.org

American Herbalists Guild
1931 Gaddis Road
Canton, Georgia 30115
(770) 751-6021
www.americanherbalistsguild.com

Herb Research Foundation
4140 15th Street
Boulder, Colorado 80304
(303) 449-2265
www.herbs.org

Northeast Herbal Association
P.O. Box 103
Manchaug, Massachusetts 01526-0103
www.northeastherbal.org

Rocky Mountain Herbalist Coalition
412 Boulder Street
Boulder, Colorado 80302
(303) 442-2215

United Plant Savers
P.O. Box 400
Barre, Vermont 05649
(802) 479-9825
www.unitedplantsavers.org

World Wildlife Fund—TRAFFIC USA
1250 24th Street, NW
Washington, DC 20037
(202) 293-4800
www.panda.org

Schools

Avena Institute
P.O. Box 333
West Rockport, Maine 04865
(207) 594-2403
www.avenainstitute.org

The Baca Institute of Ethnobotany
FLC 7195
Durango, Colorado 81301
(719) 256-4135
http://anthro.fortlewis.edu/ethnobotany/baca/

California School of Herbal Studies
P.O. Box 39
Forestville, California 95436
(707) 887-7457
www.cshs.com

Foundations in Herbal Medicine
P.O. Box 4544
NE Albuquerque, New Mexico 87196
(888) 857-1976
www.fihm.com

Northeast School of Botanical Medicine
P.O. Box 6626
Ithaca, New York 14851
(607) 539-7172
www.7song.com

Partner Earth Education Center
1525 Danby Mountain Road
Danby, Vermont 05739
(802) 293-5996
www.partnereartheducationcenter.com

Rocky Mountain Center for Botanical Studies
2639 Spruce Street
P.O. Box 19254
Boulder, Colorado 80308
(303) 442-6861
www.gotrain.com/schools/1466.htm

Sage Mountain Herbal Retreat Center
P.O. Box 420
East Barre, Vermont 05649
(802) 479-9825
www.sagemountain.com

Seeds and Root Stock of Endangered Medicinals

Abundant Life Seed Foundation
Box 772
Port Townsend, Washington 98368
(360) 385-7192
www.abundantlifeseed.org

Elixir Farm Botanicals
General Delivery
Dora, Missouri 65637
(417) 261-2393
www.elixirfarm.com

High Mowing Seed Company
813 Brook Road
Wolcott, Vermont 05680
(802) 888-1800
www.highmowingseeds.com

Horizon Herbs, LLC
P.O. Box 69
Williams, Oregon 97544
(541) 846-6704
www.horizonherbs.com

Land Reformers
35703 Loop Road
Rutland, Ohio 45775
(740) 742-3478

Publications

American Herb Association Newsletter
P.O. Box 1673
Nevada City, California 95959
(530) 265-9552
www.ahaherb.com

HerbalGram
6200 Manor Road
Austin, Texas 78723
(512) 926-4900
www.herbalgram.org

Medical Herbalism
P.O. Box 20512
Boulder, Colorado 80308
(303) 541-9552
www.medherb.com

Notes

Chapter 7

1. An article in the September-October 2001 issue of *Cooperative Grocer* (Southworth, 2001), notes the five major distributors of natural foods: Tree of Life, UNFI, Nature's Best, Blooming Prairie, and Northeast Cooperatives. By 2003, only three of these distributors independently existed after UNFI acquired Blooming Prairie and Northeast Cooperatives.

2. "Before a company runs an ad, it has to have a 'reasonable basis' for the claims. A 'reasonable basis' means objective evidence that supports the claim" (FTC, 2003).

Chapter 8

1. When I called PepsiCo to confirm that indeed *Chimaphila umbellata* is used in flavoring their cola, they told me that the recipe was proprietary and would not comment on the sources for their flavorings.

2. Black cohosh is exported to Germany in large quantities for the manufacture of Remifemin, an herbal product used to support menopause.

Chapter 9

1. The Heisenberg uncertainty principle states that the position and momentum of an electron cannot be simultaneously known with precision. The Pauli exclusion principle states that no two electrons can have identical quantum numbers within a single atom. For further discussion of these two revolutionary ideas in physics see Searway, Beichner, Jewett (2000) sections 42.5-42.6.

2. It is estimated that 180,000 people per year die due to medical error (Leape, 1994).

Chapter 10

1. Preterm infants who were massaged on a regular basis gained 47 percent more weight and were discharged six days earlier than infants without the physical contact (Field et al., 1986).

Chapter 11

1. At the time of our interview, James Green was the director of the California School of Herbal Studies (CSHS) where Leslie teaches gardening to herbal students.

2. Findhorn is an intentional community in northern Scotland that is dedicated to educating people about Earth-based spirituality. The gardens at Findhorn are world renowned.

Chapter 12

1. An inventory of wild black cohosh plants is underway. For a synopsis, see Ford (2001).

2. An herbalist-run business based in Broadway, New Jersey, Herbalist & Alchemist is owned by David Winston, a prominent herbal teacher.

3. For more details on Kate Gilday's flower essences, see <http://www.woodlandessence.com>.

4. Although there are various estimates of the number of native bison, 60 million is a generally excepted figure (Krech, 2001).

5. Berberine is an alkaloid present in goldenseal that has antibacterial properties. Berberine can be found in a number of other medicinal plants that Kate Gilday mentions.

Chapter 13

1. The Omega Center is a retreat center offering classes in natural living, spirituality, and alternative therapies located in Rhinebeck, New York.

Chapter 15

1. The quoted line is from Bob Dylan's *Subterranean Homesick Blues*.

2. David is referring to Germany's creation of the Commission E in the late 1970s. The Commission E set about to evaluate phytomedicines and establish "reasonable certainty" as to their safety and efficacy. The American Botanical Council gathered the Commission E reports on indivudal herbs and published them in a single volume. See Blumenthal et al. (1998).

3. Although precise statistics are unavailable, this is generally accepted as true (Maxey, 1999).

4. This volume has been published since I spoke with David. See *PDR* (2000).

5. Flexnor Report of 1910 sought to evaluate the existing medical schools in the United States, alternative and regular. The report, funded and conducted by those with interest in the up-and-coming biomedical model, cited alternative medical schools as inferior to the regular schools in terms of laboratory equipment and other teaching facilities. In essence, this report effectively closed down the alternative schools, making it impossible to train in alternative medicine (Griggs, 1997).

6. This quote is attributed to the prophet Muhammad.

Chapter 16

1. Estimates vary from hundreds of million of dollars to $6 billion depending on what products are included in the figure. See Brevoort, 1998.

Chapter 17

1. See current FDA opinion on ephedra at FDA, 2003a.
2. On June 27, 1997, ABC News *20/20* aired a segment titled, "Using Herb St. John's Wort to Treat Depression." Although ABCNews online does not archive segments from 1997, a transcript of the piece is available at <http://www.worldwidelabs.com/Prod_SJOHN.htm>.
3. For details on the AHG stance on certification issues see Romm (2002) and Tierra (2002).
4. Steven is referring to the Complementary and Alternative Health Care Freedom of Access Act, which gave consumers the right to choose their preferred health care modality as long as practitioners disclosed the capabilities of their therapies. The Freedom of Access act passed in Minnesota, California, and is currently before the Committee on Health and Ecology in Georgia. For details see the Complementary and Alternative Medicine Association (CAMA) Web site at <http://www.camaweb.org/> and, particularly, Roberts (1999).
5. The AHG now publishes the "Directory of Herbal Education" which describes programs across the country. Listing in this publication does not, however, indicate endorsement by the AHG. See <www.americanherbalistsguild.com>.

Chapter 18

1. For an excellent history of chiropractic medicine, see Whorton (2002).

Chapter 20

1. Annie constantly monitors the status of wild osha populations, which is listed by United Plant Savers as an at-risk medicinal due to loss of habitat and overharvesting. At the time of editing, Annie had discontinued her purchase of wild osha roots due to extraordinarily dry weather in the harvest region. She feels that the habitat is too fragile at this time to withstand a harvest. Her business, Purple Coneflower Herbals, will honor the plant by using the osha extract currently in stock.

Chapter 21

1. The 1996 creation of the Office of Alternative Medicine at the National Institutes of Health certainly points to the growth in inquiry. This office's budget increased from $7.4 million in 1996 to $12 million in 1997 (NIH, 1997). In 1999, this office further grew into the National Center for Complementary and Alternative medicine with a budget of $50 million (Walker, 2000). For further discussion on re-

search into complementary and alternative medicines, see Milbank Memorial Fund (1998).

2. For an explanation of DSHEA, see CFSAN, 1995.

3. I believe that Michael is referring to regulations set up by the German government's Commission E, which allowed use of herbal medicines if they could be proved safe with "reasonable certainty." See Blumenthal et al. (1998). In Germany, if herbal medicines are used under the guidance of a physician, the cost can be covered under the national health care plan (Berman et al., 1992).

4. The survey that Michael discusses has since been completed. For further information, contact the American Herbal Products Association.

Chapter 22

1. David refers to the Charter of Rights for Herbalists, which was created during the reign of Henry VIII to quiet troubles between surgeons and physicians, the professional brand of herbalist in that day. It says,

> it shall be lawful to every person being the King's subject, having knowledge and experience of the nature of Herbs, Roots and Waters, or of the operation of same, by speculation of practice within any part . . . of the King's Dominions, to practice, use and minister in and to any outward sore, uncome, wound, apostumations, outward swelling or disease, according to their cunning, experience and knowledge in any of the diseases, sores and maladies beforesaid, and all other like to the same, or drinks for the Stone and Strangury, or Agues, without suit, vexation, trouble, penalty, or loss of their goods. (Griggs, 1997, p. 59)

Modern herbalists in England use the charter as the legal basis for clinical practice (Griggs, 1997).

2. Although Mao Tse-tung first outlawed the practice of traditional Chinese medicine (calling it shamanic and superstitious) and instead concentrated on developing the practice of Western medicine in China, in 1954 he formed the Department of Chinese Medicine. In doing so, Mao hoped to add a distinguishing feature to his brand of communism and demonstrate a sense of Chinese self-reliance and independence. Through the Department of Chinese Medicine, a new brand of practice which fused traditional Chinese healing techniques with Western medicine was formed. Several state-sponsored colleges were also established to teach this particular version of traditional Chinese medicine.

3. The idea of "protein complementing" was widely promoted by Frances Lappe (1985) in her book, *Diet for a Small Planet*. Research that supports this theory was done on rats who have vastly different nutritional requirements than humans. Since then nutritional studies have been done which suggest essential amino acids can be gotten from single food sources as long as they are unrefined (Young and Pellet, 1994).

Chapter 23

1. The Biotechnology Council of New Jersey lures researchers to the area by saying that seventeen out of the twenty largest pharmaceutical houses have major facilities in the state.

Chapter 24

1. Two beetles have been introduced to eradicate St. John's wort, *Agrilus hyperici* (McCaffrey et al., 1995) and *Chrysolina quadrigemina* (Huffaker and Kennett, 1959; CDFA, 2004).
2. The Nova Scotia Weed Control Act lists fifteen plants including coltsfoot that are considered class-one weeds. This means that they can spread from the wild easily and can endanger adjacent pastureland. For more information visit the Nova Scotia Department of Agriculture Web site at <http://www.gov.ns.ca/nsaf/rir/weeds/aboutact.htm>.
3. Herbs in health food products include Odwalla's Wellness Juice with Echinacea and Astragalus (Odwalla, 2003) and Fresh Samantha's Super Juice with Echinacea (Fresh Samantha, 2003).
4. The domino effect refers to the position of one plant on the endangered plant lists followed closely by the recommended substitutes. Herbalists are working hard to eliminate the use of certain plants in their practices, but it seems that they are only endangering other plants in the process. See Chapter 5 for further discussion.
5. Berberine, one the constituents in goldenseal, is touted for its antibacterial properties. Berberine is also present in Oregon grape root and gold thread, which are popular substitutes in the herbal materia medica. All three plants are listed by United Plant Savers as at risk or to watch.
6. John Robbins is the author of *The Food Revolution* (2001) and *Diet for a New America* (1987).

Chapter 25

1. For details on GMPs, see *FDA* (2002, 2003b).
2. For details on DSHEA, see CFSAN (1995).

Chapter 26

1. Peeka is referring to the school now known as the North American College of Botanical Medicine.
2. The Green Nations Gathering, held each fall in the Catskill region of New York State, is a conference aimed at the grassroots, community-centered herbalist.
3. 7song, a native of New York City, has for the past decade lived and taught Earth-centered herbalism in the Finger Lakes region of New York State. His teachings are based on a fusion of Chinese, Ayurvedic, and eclectic herbal therapeutics.

Bibliography

Alland, A. (1970). *Adaptations in cultural evolution: An approach to medical anthropology.* New York: Columbia University Press.

American Herbal Products Association (AHPA) (2003). American Herbal Products Association home page. Available from <http://www.ahpa.org>.

American Herbalists Guild (AHG) (2003). American Herbalists Guild home page. Available from <http://www.americanherbalistsguild.com>.

Anaheim Goldenseal Planning Consortium (1998). Industry and government form partnership for goldenseal conservation. *United Plant Savers Newsletter* 1(2): 1,3. Available from United Plant Savers, P.O. Box 400, East Barre, VT 05649.

Angell M. and Kassirer J. (1998). Alternative medicine—The risks of untested and unregulated remedies [editorial]. *New England Journal of Medicine* 339(12): 839-841.

Anonymous (1998). Herb sales up 101 percent in mainstream market. *HerbalGram* 43: 61.

Arthur Andersen LLP (1999). 1998 goldenseal survey results, March. Silver Spring, MD: American Herbal Products Association.

Arvigo, R. (1994). *Sastun: My apprenticeship with a Maya healer.* San Francisco: HarperCollins.

Balick, M. and Cox, P. (1996). *People, plants, and culture: The science of ethnobotany.* New York: Scientific American Library.

Banks, A. (1991). *First person America.* New York: W.W. Norton and Company.

Barnes, P., Powell-Griner, E., McFann, K., and Nahin, R. (2004). Complementary and alternative medicine use among adults: United States, 2002. CDC Advance Data Report 343, May 27.

Batmanghelidg, F. (1995). *Your body's many cries for water.* Berwyn, PA: Global Health Solutions.

Berman, B., Larson, D., and Swyers, J. (1992). Alternative medicine, expanding medical horizons. Report on alternative medical systems and practices in the United States. Washington DC: NIH. Available from <http://www.naturalhealthvillage.com/reports/rpt2oam/toc.htm>.

Berry, T. (1988). *The dream of the earth.* San Francisco: Sierra Club Books.

Biotechnology Council of New Jersey (2003). New Jersey Biotechnology Online. Available from <http://www.newjerseybiotech.org/career/destination_industry.htm>.

Bliss, S. (1997). *Seeds of herbalism.* Stowe, VT: Rooted Wisdom Center for Wholistic Studies.

Blumenthal, M. (2000). Goldenseal. In Gladstar, R. and Hirsch, P. (Eds.), *Planting the future: Saving our medicinal herbs* (pp. 111-122). Rochester, VT: Healing Arts Press.

———. (2002). Herb sales down in mainstream market, up in natural food stores. *HerbalGram* 55: 60.

———. (2003). Herbs continue slide in mainstream market: Sales down 14 percent. *HerbalGram* 58: 71.

Blumenthal, M., Busse, W., Klein, J., Klein, S., Riggins, C., and Rister, R. (1998). *The complete German Commission E monographs: Therapeutic guide to herbal medicines.* Austin, TX: American Botanical Council.

Bradford, H. (2002). White House Commission on Complementary and Alternative Medicine Policy issues final report. *HerbalGram* 56: 10-11.

Brevoort, P. (1998). The booming U.S. herb market. *HerbalGram* 44: 33-48.

Brody, J. (1999). Americans gamble on herbs as medicines. *The New York Times*, February 9, 1999, p. D1.

Buhner, S. (1996). *Sacred plant medicine: Explorations in the practice of indigenous herbalism.* Coeur d'Alene, ID: Raven Press.

———. (2002). *The lost language of plants: The ecological importance of plant medicines to life on Earth.* White River Junction, VT: Chelsea Green Publishing.

California Department of Food and Agriculture (CDFA) (2004). California Department of Food and Agriculture home page. Available from <http://www.cdfa.ca.gov/index.htm>.

Capit, W. and Elson, L. (2001). *The anatomy coloring book.* Upper Saddle River, NJ: Pearson Educational.

Capra, F. (1975). *The tao of physics.* New York: Bantam Books.

———. (1983). *The turning point.* New York: Bantam Books.

———. (1996). *The web of life.* New York: Anchor Books.

Cech, R. (2000). *Making plant medicine.* Williams, OR: Horizon Herbs.

———. (2000). UpS "at risk" forum: Response to letter from Becca Harber. *Journal of Medicinal Plant Conservation* 2(2): 17.

———. (2002). *Growing at-risk medicinal herbs.* Williams, OR: Horizon Herbs.

Center for Food Safety and Applied Nutrition (US) [CFSAN] (1995). Dietary Supplement Health and Education Act of 1994. Washington DC: Food and Drug Administration. Available from <http://vm.cfsan.fda.gov/~dms/dietsupp.html>.

Center for Food Safety and Applied Nutrition (US) [CFSAN] Industry Activities Branch (1999). Guidance for industry: Statement of identity, nutrition labeling, and ingredient labeling of dietary supplements, small entity compliance guide. Washington, DC: Food and Drug Administration (US) [FDA]. Available from <http://vm.cfsan.fda.gov/~dms/ds-label.html>.

Cheeke, P. (1998). *Natural toxicants in feeds, forages and poisonous plants.* Danville, IL: Interstate Publishers.

Chester, P. (1998). *Sisters on a journey.* New Jersey: Rutgers University Press.

Complementary and Alternative Medicine Association (CAMA) (2003). CAMA home page. Available from <http://www.camaweb.org/>.
Cotton, C. (1996). *Ethnobotany: Principles and applications.* Chinchester, UK: John Wiley and Sons.
Cowan, E. (1995). *Plant spirit medicine.* Mill Spring, NC: Swan, Raven and Company.
Davis-Floyd, R. and St. John, G. (1998). *From doctor to healer: The transformative journey.* New Brunswick, NJ: Rutgers University Press.
Eisenberg, D., Davis, R., Ettner, S., Appel, S., Wilkey, S., Van Rompay, M., and Kessler, R. (1998). Trends in alternative medicine use in the United States, 1990-1997. *Journal of the American Medical Association* 280(18): 1569-1575.
Eskinazi, D. and Muehsam, D. (2000). Factors that shape alternative medicine: The role of the alternative medicine research community. *Alternative Therapies* 6(1): 49.
Federal Trade Commission (US) [FTC] (2003). Frequently asked advertising questions: A guide for small business: General advertising policies. Available from <http:// www.ftc.gov/bcp/conline/pubs/ad-faqs.htm>.
Field, T., Schanberg, S., Scafidi, F., Bauer, C., Vega-Lahr, N., Garcia, R., Nystrom, J., and Kuhn, C. (1986). Tactile/kinesthetic stimulation effects on preterm neonates. *Pediatrics* 77: 654-658.
Findhorn Foundation (2003). Findhorn Foundation home page. Available from <http://findhorn.org/>.
Food and Drug Administration (US) [FDA] (2002). Current good manufacturing practices in manufacturing, packing, or holding human food. [monograph online] Available from <http://vm.cfsan.fda.gov/~lrd/cfr110.html>.
Food and Drug Administration (US) [FDA] (2003a). Evidence on the safety and effectiveness of ephedra: Implications for regulation. [monograph online] Available from <http://www.fda.gov/bbs/topics/NEWS/ephedra/whitepaper.html>.
Food and Drug Administration (US) [FDA] (2003b). FDA proposes labeling and manufacturing standards for all dietary supplements. [monograph online] Available from <http://www.fda.gov/bbs/topics/NEWS/2003/NEW00876.html>.
Ford, P. (2001). Pilot inventory study of black cohosh. *Journal of Medicinal Plant Conservation* Spring 2001: 23.
Ford, R. (1987). Ethnobotany: Historical diversity and synthesis. In Ford, R. (Ed.), *The nature and status of ethnobotany* (pp. 33-49). Ann Arbor, MI: University of Michigan.
Foster, S. and Chongxi, Y. (1992). *Herbal emissaries: Bringing Chinese herbs to the West.* Rochester, VT: Healing Arts Press.
Fox, M. (1983). *Original blessing.* Santa Fe, NM: Bear and Company.
Frawley, D. (1992). Herbs and the mind. In Tierra, M. (Ed.), *American herbalism* (pp. 144-157). Freedom, CA: The Crossing Press.
Fresh Samantha Juices (2003). Fresh Samantha Healthy Sips page. Available from <http://www.freshsamantha.com/healthysips/juices/super.htm>.

Gerard, J. (1975). *The herbal or general history of plants.* New York: Dover Publishing.

Gladstar, R. (1993). *Herbal healing for women.* New York: Simon and Schuster.

———. (2001). *Rosemary Gladstar's family herbal: A guide to living life with energy, health, and vitality.* North Adams, MA: Storey Books.

Gladstar, R. and Hirsch, P. (2000). *Planting the future: Saving our medicinal herbs.* Rochester, VT: Healing Arts Press.

Green, J. (1990). Introduction to folkloric herbalism. In Tierra, M. (Ed.), *American herbalism* (pp. 37-40). Freedom, CA: The Crossing Press.

———. (1991). *The male herbal: Health care for men and boys.* Freedom, CA: The Crossing Press.

———. (2000). *The herbal medicine maker's handbook: A home manual.* Freedom, CA: The Crossing Press.

Green Hope Farm (2002). Green Hope Farm home page. Available from <http://www.unwindings.com/greenhopefarm.htm>.

Greenwald, J. (1998). The herbal medicine boom. *Time* 152(21): 61.

Grele, R. (Ed.) (1975). *Envelopes of sound.* Chicago, IL: Precedent Publishing.

Griggs, B. (1997). *Green pharmacy: The history and evolution of Western herbal medicine.* Rochester, VT: Healing Arts Press.

Grossinger, R. (1995). *Planet medicine: Origins.* Berkeley, CA: North Atlantic Books.

Haley, A. (1976). *Roots.* Garden City, NY: Doubleday and Company.

Harpignies, J.P. (1995). The greening of medicine. *Lapis* 1: 61-65.

Heisenberg, W. (1958). *Physics and philosophy.* New York: Harper Torchbooks.

Hoffman, D. (1992). *Herbalism: Gaia in action in therapeutic herbalism.* California: self-published.

———. (1997). *The new holistic herbal.* Rockport, MA: Element Books.

———. (2003). *Medical herbalism: Principles and practices.* Rochester, VT: Healing Arts Press.

Hopman, E. and Bond, L. (Eds.) (1996). *People of the earth: The new pagans speak out.* Rochester, VT: Destiny Books.

Horne, S. (1995). *The ABC herbal: A simplified guide to natural health care for children.* Winona Lake, IN: Whitman Books.

Huffaker, C. and Kennett, C. (1959). A ten-year study of vegetation change associated with biological control of Klamath weed. *Journal of Range Management* 12: 69-82.

Illich, I. (1977). *Medical nemesis: The expropriation of health.* New York: Bantam Books.

Jensen, D. (2001). Saving the indigenous soul: An interview with Martin Prechtel. Available from <http://hiddenwine.com/indexSUN.html>.

Johnston, B. (1997). One-third of nation's adults use herbal medicines. *HerbalGram* 40: 49.

Kirkland, J., Matthews, H., Sullivan, C., and Baldwin, K. (Eds.) (1992). *Herbal and magical medicine*. Durham, NC: Duke University Press.

Krech, S. (2001). Buffalo tales: The near-extermination of American bison. [monograph online] Available from <http://www.nhc.rtp.nc.us:8080/tserve/nattrans/ntecoindian/essays/buffalo.htm>.

Landis, R. and Khalsa, K.P. (1997). *Herbal defense: Positioning yourself to triumph over illness and aging*. New York: Warner Books.

Landy, D. (1977). *Culture, disease, and healing: Studies in medical anthropology*. New York: Macmillan.

Lappe, F. (1985). *Diet for a small planet: Twentieth anniversary edition*. New York: Ballantine Books.

Leape, L. (1994). Error in medicine. *Journal of the American Medical Association* 272(23): 1851-1857.

Liebmann, R. (1997). Planting the future. *United Plant Savers Newsletter* 1(2): 6-8.

Losey, J., Rayor, L., and Carter, M. (1999). Transgenic pollen harms monarch larvae. *Nature* 399: 214.

Low Dog, T. (1995). An informal survey of herb usage amongst HIV patients in New Mexico. Presented during Ambulatory HIV Rotation at UNM Hospital.

———. (1996). Unpublished interview with Suzanne Richman at Sage Mountain Herbal Retreat Center.

———. (1997). *Foundations in herbal medicine*. Albuquerque, NM: Self-published.

Martin, G. (1995). *Ethnobotany*. Cambridge, UK: University Press.

Maxey, A. (1999). The dangers of herbal supplements: What the consumer needs to know. [monograph online] Available from <http://www.du/edu/~amaxey/casebook.htmll>.

McCaffrey, J., Campbell, C., and Andres, L. (1995). St Johns-wort *Hypericum perforatum* L. (Clusiaceae). In Nechols, J., Andres, L., Beardsley, J., Goeden, R., and Jackson, C. (Eds.), *Biological control in the U.S. Western Region: Accomplishments and benefits of Regional Research Project W-84, 1964-1989* (pp. 281-285). Oakland: University of California, Division of Agriculture and Natural Resources, Publication 3361.

Medline Plus Health Information (2003). Medline Plus home page. Available from <http://medlineplus.gov/>.

Messengers of the Gods: Conversations with traditional healers of Belize [videocassette] (1995). New York: New York Botanical Gardens. 1 videocassette. 70 mins. Out of print.

Milbank Memorial Fund (1998). Enhancing the accountability of alternative medicine. [monograph online] Available from <http://www.milbank.org/mraltmed.html#rl>.

Mills, S. (1991). *Out of the earth: The essential book of herbal medicine*. Middlesex, UK: Viking Arkana.

Moss, W. (1974). *Oral history program manual*. New York: Praeger Publishing.

Narby, J. (1998). *The cosmic serpent: DNA and the origins of knowledge.* New York: Jeremy P. Tarcher/Putnam.

National Institutes of Health [NIH] (1997). Complementary and Alternative Medicine at the NIH. *Alternative Medicine Clearing House* April 4: 1.

National Institute of Medical Herbalists [NIMH] (2003). National Institute of Medcial Herbalists home page. Available from <http://www.nimh.org.uk/>.

Neas, J. (2003). Native Prairie and KDWP. [monograph online] Available from <http://skyways.lib.ks.us/orgs/jayhawkaudubon/nativeprairie.html>.

Neel, R. and Bloom, M. (1996). No Title. *Physician's Weekly* 13(28).

Neuenschwander, J. (1976). *Oral history as a teaching approach.* Washington, DC: National Educational Association.

Newall, C., Anderson, L., and Phillipson, J. (1996). *Herbal medicines: A guide for health-care professionals.* London: The Pharmaceutical Press.

Ni, M. (Translator) (1995). *The yellow emperor's classic of medicine.* Boston: Shambhala Publications.

North American College of Botanical Medicine (2002). North American College of Botanical Medicine home page. Available from <http://www.swcp.com/botanicalmedicine/>.

Nova Scotia Department of Agriculture and Fisheries (CA) [NSAF] (2003). About the Nova Scotia Weed Control Act. Available from <http://www.gov.ns.ca/nsaf/rir/weeds/aboutact.htm>.

Odwalla, Inc. (2003). Odwalla product information page. Available from <http://www.odwalla.com/flavor/ensemble.html>.

Omega Center (2003). Omega Center home page. Available from <http://www.eomega.org/>.

PDR for Herbal Medicines, Second Edition (2000). Montvale, NJ: Medical Economics.

Prechtel, M. (1998). *Secrets of the talking jaguar.* New York: Jeremy P. Tarcher/Putnam.

———. (1999). *Long life, honey in the heart: A story of initiation and eloquence from the shores of a Mayan lake.* New York: Jeremy P. Tarcher/Putnam.

Robbins, J. (1987). *Diet for a new America.* Walpole, NH: Stillpoint.

———. (2001). *The food revolution: How your diet can help save your life and our world.* Berkeley, CA: Conari Press.

Roberts, M. (1999). Frogwater and you [monograph online]. Available from <http://www.camaweb.org/library/public_policy/frogwater.php>.

Robin, K. (2002). Small minority accounts for majority of botanical product sales. *HerbalGram* 56: 53.

Romanucci-Ross, L, Moerman, D., and Tan Prechtel, M. (1999). *Long life, honey in the heart: A story of initiation and eloquence from the shores of a Mayan lake.* New York: Jeremy P. Tarcher/Putnam.

Romm, A. (2002). Responses on education and certification from the American Herbalists Guild (AHG) for the White House Commission on Complementary and Alternative Medicine Policy. [monograph online] Available from <http://www.americanherbalistsguild.com>.

Salmone, E. (1992). The marginalization of indigenous ethnobiological knowledge as a result of foreign contact. Originally presented at the Society for Ethnobiology Conference, March 1. Washington, DC.

Searway, R., Beichner, R., and Jewett, J. (2000). *Physics for scientists and engineers, with modern physics,* Fifth edition. Philidelphia, PA: Saunders College Publishing.

Shen-nong Limited (2002). Modern China 1912 A.D. - Present [monograph online]. Available from <http://www.shen-nong.com/eng/shen-nong/history/modern/modern.htm>.

Singer, P. (Ed.) (1977). *Traditional healing: New science or new colonialism.* North Scituate, MA: Duxbury Press.

Skinner, W. (1999). ASP Conference reviews science and regulations. *HerbalGram* 47: 68-70.

Soule, D. (1995). *The roots of healing: A woman's book of herbs.* Secaucus, NJ: Carol Publishing Group.

Southworth, G. (2001). Natural/organic industry outlook. *Cooperative Grocer* 96: 23.

Stapp, H. (1971). S-matrix interpretation of quantum theory. *Physical Review* Volume D3 (March 15): 1310.

Starhawk (1989). *The spiral dance: A rebirth of the ancient religion of the great goddess.* San Francisco: Harper and Row.

Starr, P. (1982). *The social transformation of American medicine.* New York: Basic Books, Inc.

Stenn, F. (1980). Thoughts of a dying physician. *Forum on Medicine* 3: 719.

Thompson, P. (1978). *The voice of the past.* Oxford, UK: Oxford University Press.

Tierra, L. (2002). AHG certification and legislative issues [monograph online]. Available from <http://www.americanherbalistsguild.com>.

Tierra, M. (1990). *The way of herbs,* Third edition. New York: Pocket Books.

———. (Ed.) (1992). *American herbalism.* Freedom, CA: The Crossing Press.

Tilgner, S. (1991). *Edible and medicinal herbs, volume I* [videocassette]. Sammamish, WA: Tree Farm Communications.

———. (1993). *Edible and medicinal herbs, volumes I and II* [videocassette]. Sammamish, WA: Tree Farm Communications.

———. (1999). *Herbal medicine from the heart of the Earth.* Pleasant Hill, OR: Wise Acres Publishing.

TRAFFIC (2003). TRAFFIC home page. Available from <http://www.traffic.org/>.

United Plant Savers [UpS]. Educational Material Presentation. (1999 June). Available from United Plant Savers, PO Box 400, East Barre, VT 05649.

Walker, L. (2000). Mysticism and/or rigor: Can science and alternative medicine shake hands? [monograph online] Available from: <http://www.columbia.edu/cu/21stC/issue-3.4/walker.html>.

Weed, S. (1992). *Menopausal years, the Wise Woman way: Alternative approaches for women 30 to 90*. Woodstock, NY: Ash Tree Publishing.

Weil, A. (1995). *Spontaneous healing*. New York: Fawcett Columbine.

Whorton, J. (2002). *Nature cures: The history of alternative medicine in America*. New York: Oxford University Press.

Willett, W., Giovannucci, E., Rimm, E., Stampfer, M., Colditz, G., and Ascherio, A. (1990). Relation of meat, fat and fiber intake to the risk of colon cancer in a prospective study among women. *New England Journal of Medicine* 323(24): 1664-1672.

Wolfson, P. and Hoffmann, D. (2003). An investigation into the efficacy of *Scutellaria lateriflora* in healthy volunteers. *Alternative Therapies in Health and Medicine* 9(1): 74-78.

Wood, M. (1997). *The book of herbal wisdom*. Berkeley, CA: North Atlantic Books.

———. (2000). *Vitalism: The history of herbalism, homeopathy, and flower essences*. Berkeley, CA: North Atlantic Books.

Woodland Essences (2003). Woodland Essences home page. Available from <http://www.woodlandessence.com/>.

World Wildlife Federation (US) [WWF] TRAFFIC USA. (1999). Comparative analysis of management regimes and medicinal plant trade monitoring mechanisms for American Ginseng and Goldenseal. *Conservation Biology* 14(5): 1422-1434.

Worsley Classical Five Element Acupuncture (2003). J.R. Worsley home page. Available from <http://www.worsleyacupuncture.com/>.

Young, V. and Pellett, P. (1994). Plant proteins in relation to human protein and amino acid nutrition. *American Journal of Clinical Nutrition* 59(5): 1203S-1212S.

Yow, V. (1994). *Recording oral history: A practical guide for social scientists*. London: Sage Publishing.

Index

The ABC Herbal: A Simplified Guide to Natural Health Care for Children, 227, 228
ABC Herbs, 227
Accreditation, 51-52, 203-204, 261-262
Achillea millefolium (yarrow), 40, 90, 233
Aconite, 52, 242
Acupuncture
 choice of healing methods, 145
 licensure, 243, 246-247, 284-286
 limitations of treatment, 250
Adrenal burnout, 236
Aesthetics of herbal medicine, 11, 18, 189
AHG. *See* American Herbalists Guild
AHPA. *See* American Herbal Products Association
Alien species in habitat loss, 301-302
Allies, plant, 156
Allopathic medicine
 defined, 320
 holistic healing versus, 138, 143, 269, 289, 305-306, 319-320
 license to kill, 133-134
 move to botanical medicine, 15
ALOHA (Aware of Life Options and Healing Arts), 291
Alternative health care, 260-261
Alternative plant use, 163
AMA. *See* American Medical Association
American barberry, 168
American Cancer Society report on diet and cancer, 223
American Ginseng. *See* Ginseng
American herbal community, 6-8

American Herbal Products Association (AHPA)
 coordination with U.S. Fish and Wildlife Service, 280
 industry standards, 313-314
 president Michael McGuffin, 63, 273
 purpose, 153
American herbal renaissance
 beginning of movement, 73
 cyclical nature of, 152-153
 history, 83, 186
 overview, 30
American herbalism
 borrowing, 89-90
 changes, 204
 diversity of approaches, 13-14
 essence of, 200-202
 European versus, 241-242
 expansion and potential, 209-210
 history, 185-186, 199-200
 introduction of science, 203-204
 product of culture, 88
American Herbalists Guild (AHG)
 agent for health providers, 293
 certification not required, 58
 director Steven Horne, 227
 educational guidelines, 230, 232
 folkloric herbalism, 41
 history, 213
 licensure, 49-50, 56, 230-231
 member David Milbradt, 283
 member Peeka Trenkle, 319
 past president David Hoffmann, 199
 professionalizing vision, 251-252
 purpose, 9, 153
 training store employees, 194
 use of saw palmetto, 223
 values, 298

American Medical Association (AMA)
   herbalists as members, 205
   herbs versus herbalism, 207
   language of science, 208
   marketplace therapeutics, 193
Amino acids, 287
The Anatomy Coloring Book, 221-222
Angelica root, 147
Anger, impact on liver, 235
Animal dissection for education, 128-129
Antibacterial, resistance by bacteria, 288-289
Antibiotics and herbs, 134-135, 250-251
Antimicrobial actions of goldenseal, 75
Antispasmodic action of cyanogenic glycosides, 90
Anxiety, kava as remedy, 15
Apprenticeship, 179-181, 272
Aristotle, 34
Arnica, 242, 245
Art of Birthing, 319
Arvigo, Rosita, 58, 91, 92-93, 95
Aspirin, 258
Astragalus, 147
Atkins diet, 141
At-risk plants
   contributing factors, 113-115
   criteria for listing, 115-116
   harvesting, 161-162
   listing, 333
Avena Botanicals, 299, 303
Avena Institute, 299
Aware of Life Options and Healing Arts (ALOHA), 291
Ayurveda, 213, 243

Babineau, Don, 161
Back-to-the-land movement, 73, 152, 174, 186
Bacteria resistance to antibiotics, 288-289
Balance, holistic concept, 274, 306
Balick, Michael, 91
Barberry, 168
Bastyr, 203

Batcha, Laura, 99-111
Belief and success of treatment, 139-140
Belize Association of Traditional Healers, 91
Belladonna
   danger of use, 52, 242
   restricted use, 53, 241
   side effects, 132
Berberine
   domino effect with endangered plants, 304
   goldenseal, 75, 76
   plants other than goldenseal, 168
Berry, Thomas, 41, 62-63
Bhajan, Yogi, 238
Biodynamically grown herbs, 312
Biomedicine
   enhancement of herbalism, 35-39
   science of, 34-35
   understanding of herbalism, 47-48
Bioregional herbalism
   booklet template, 209
   consciousness of, 196
   defined, 7
   endangered plants, 300-301
   healing approach, 13
   solution to domino effect, 77
   solution to monopopularizing, 120
Black cohosh
   CITES impact, 119
   cultivated product, 66
   endangered plants, 187, 281, 300, 302, 312
   export to Europe, 241-242
   harvesting, 162
   market popularity, 63
   monopopularizing, 70, 196
   organic growing, 7, 146
   replanting, 164
   substantiation of claims, 282
   uses of, 15, 147
Blood regulation and hawthorn berries, 40
Blue cohosh
   endangered plants, 300
   harvesting, 80, 163

Blue cohosh *(continued)*
   monopopularizing, 196
   organic growing, 146
   uses of, 147
Body-mind-spirit connection, 17-18, 26-28, 192
Borrowing American herbalism, 89-91
Botanical medicine
   biomedicine and physics comparison, 35
   defined, 12, 33
   Earth connection, 18-20
   mind-body connection, 17-18
   overview, 14-15
   scientific evaluation of herbs, 16-17
   specific herbs for specific healing, 15
Bryonia *(Bryonia alba)*, 312
BT corn and the monarch butterfly, 303-304
Buffalo herd reduction, 167
Buhner, Stephen, 41-42, 61-62, 79
Burdock, 271

Cajun herbalism, 202
Calamus root, 300
Calendula, 120
California poppy, 37
California School of Herbal Studies (CSHS)
   classroom learning, 158
   director David Hoffmann, 199
   former director James Green, 185
   founder Rosemary Gladstar, 173, 200-201
   Garden Apprentice Program, 149
   purpose, 186, 187
   reciprocity of plants, 96
   science and folklore, 159
Calling of herbs, 3-4, 46
CALM (Center for Alternative Living Medicine), 283
CAM (complementary and alternative medicine), 50-51, 53

Canadian herbalism, 125, 278
Capra, Fritjof, 43, 44, 94
Cartesian worldview
   American herbalism, 87-88
   anxiety of, 94
   biomedical roots, 47
   healing process, 43
   herbalism certification, 58-59
   isolationist attitude, 90
   model of reality, 45
   separating body and spirit, 234
   short-sightedness, 77
   substitute products, 76
Cascara sagrada, 236
Cayenne, 236
CDC (Centers for Disease Control and Prevention), 55
Cech, Richo
   at-risk plants, 113-117
   background, 113
   complications of herbalism study, 71
   corporate harvesting, 67
   goldenseal listing, 117-119
   monopopularizing, 70, 119-120
   plant conservation, 66-67
   preservation of prairie, 61
   unity of herbalism, 57
   UpS membership, 120-121
Center for Alternative Living Medicine (CALM), 283
Centers for Disease Control and Prevention (CDC), 55
Centrum sales, 311
Ceremony renewal of connection, 41-42
Certification. *See also* Licensure
   defined, 50
   efficacy, 106-107
   exams, 286
   levels, 49
   overview, 49-50
   reservations, 57
Chamaelirium luteum (False unicorn), 116
Chamomile, 105, 226
Chaparral, 277

Chemical poisoning, 235-236
Chester, Penfield, 8
Chimaphila umbellata (pipsissewa), 66-67, 114
Chinese medicine
　acupuncture. *See* Acupuncture
　Ayurvedic herbalists, 58
　chi, 234
　clinical trials, 234
　formula for deficiency of fire, 236
　framework of tradition, 289
　grief connection to respiratory problems, 235
　herbology, 149
　history, 285, 294
　traditional (TCM), 149, 213, 285-286
　use with Western herbs, 295-296
Chiropractic licensure, 247, 284-285
Choice in holistic healing, 138
Cholesterol, garlic as remedy, 15
Christopher, Dr., 200
Circular thinking, 59
CITES (Convention on International Trade in Endangered Species), 117-119
Claims of products, 107
Classroom learning. *See also* Education
　apprenticeship, 179-181, 272
　benefits of, 158, 175-177
　large classes, 179-181
Classroom teaching. *See also* Education
　apprenticeship, 179-181, 272
　benefits of, 158
　communicating with plants, 151
　knowledge of teachers, 175-177
Cleansing the body, 221
Cleavers, 107
Clinical trials
　Chinese, 234
　doctor approved, 187
　herbs versus drugs, 288-289
　nature of, 108-109
　predictability of herbs, 236
Clinical versus lay herbalists, 124-126
Clinician-patient relationship, 129-131
Closed system, 15

Coho salmon, 118
Coltsfoot, 74, 289, 301, 329
Commercialization of herbs, 151-152
Communication
　among herbalists, 41
　language of science, 36-37, 208-209
　with plants, 150-151, 156
Community garden project, 262-264
Company size
　consumer expectations, 104-105
　corporatization of herbalism, 189-190
Competency and licensure, 230-232, 261, 284
Competition of products, 102
Complementary and alternative medicine (CAM), 50-51, 53
Computer
　global connections, 85
　HerbNET program, 126
Connection through ceremony, 41-42
Conscious power, 5-6, 330, 331
Conscious transplanting, 155-156
Conservation of forests, 214-215
Constitutional model, 228
Constructive ignorance, 132
Consumerism, 74, 77, 323-324
Consumers
　education, 110-111, 181-183, 247-249
　expectations of large companies, 104-105
　responsibilities, 275-276, 277-278
Convention on International Trade in Endangered Species (CITES), 117-119
Corporatization of American herbalism, 189-190
Cowan, Eliot, 93, 177
Crataegus oxycantha (hawthorn), 40, 90, 108
CSHS. *See* California School of Herbal Studies
Culpeper, 208
Cultivated herbs, 65-66, 188
Curanderos, 144
Curanderos licensure, 261-262
Cyanogenic glycosides, 90

Dandelion
    abundance of, 188, 196
    ease of growing, 211
    hangnail cure, 132
    Lasix replacement, 143
    liver detoxifier, 297
    monopopularizing, 73
    wildcrafting, 271
Davis-Floyd, Robbie, 12, 13, 14, 19-20, 29
de Bairacli Levy, Juliette, 303
Deadly nightshade. *See* Belladonna
Decentralization, 59-60
Deconstruction of American herbalism, 204
Deficiency of fire, 236
Demulcent property, 304
Depression, St. John's wort as remedy, 15, 228
Descartes, 34
Diagnosis of disease, 222
Dietary Supplement Health and Education Act (DSHEA)
    claim substantiation, 107
    consumer information, 275
    grandfathered dietary ingredients, 281-282
    labeling laws, 203
    labeling requirements, 314-315
    regulation exemptions, 102-103
Diets, 141
Dioscorides, 40
Disagreement in herbal community, 326-327
Disconnection
    from native land, 85-86, 166-167
    from Spirit, 268-269
Distribution of products, 102
DNA communication, 46
Doctor to the self, 23-24. *See also* Self-responsibility for healing
Doctrine of signatures, 168
Domino effect
    at-risk medicinals, 168
    defined, 74
    endangered plants, 304

Domino effect *(continued)*
    solution, 76-77
    substitution of herbs, 74-78, 168-169
Dong quai, 147
Dormant seed phase, 18-20
The Dream of the Earth, 41
Drugs for healing, 15, 183
Druidism, 217
Drum, Ryan
    background, 123, 124-125
    clinical versus lay herbalists, 124-126
    clinician-patient relationship, 129-131
    consumerism, 74, 77
    herbs and antibiotics, 134-135
    license to kill, 133-134
    marketplace education, 126
    popularity of herbalism, 123-124
    practitioner education, 126-129
    science and folklore, 132-133
    self-responsibility of healing, 24
    specific herbs for specific healing, 15
    understanding of herbalism, 83
DSHEA. *See* Dietary Supplement Health and Education Act
Dylan, Bob, 201

Earth connection
    dormant seed phase, 18-20
    holistic herbalism, 21-22
Echinacea *(Echinacea angustifolia)*
    at-risk listing criteria, 116
    clinical trials, 109
    cold remedy, 87, 239
    complexity of, 158
    efficacy certification, 107
    healing herb, 270
    market popularity, 63
    misuse, 52
    monopopularizing, 70, 71, 242
    natural prairie growth, 61
    organic growing, 65, 271, 279-280
    overuse of, 302
    Purple Coneflower Herbals, 267, 270
    self-medication, 241

Echinacea *(continued)*
 shortages, 65
 supply and demand, 64
 use by the general public, 52-53, 147
Echinacea angustifolia. See Echinacea
Eclectic herbal healing
 background, 13, 25, 233
 diversity of healing approaches, 58
 licensure, 213
 sharing knowledge, 209
Ecology
 domino effect, 74-78
 education, 183-184
 industry and endangered plant crisis, 63-69
 monopopularizing, 70-74
 overview, 61-63
 preservation, 242. *See also* Habitat reduction
 reality of endangered plant, 78-81
Edible and Medicinal Herbs, 309
Education
 alternative plant use, 163
 animal dissection, 128-129
 apprenticeship, 179-181, 272
 bioregional consciousness, 196
 classroom, 151, 158, 175-177, 179-181, 272
 consumer, 110-111, 239-240, 247-249, 275-276
 Elements of Good Health class, 137-139
 environment and ethnobotanical concepts, 183-184
 experience versus lecture, 179
 experiential learning, 173-174
 future of herbalism, 324-325
 gardening classes, 158-159
 herbalists, 230, 259, 321-322
 licensure of school graduates, 51-52
 marketplace, 126, 141
 master herbalists and Wise Woman, 157-158
 practitioner, 126-129, 145-146
 prevention of overharvesting, 271

Education *(continued)*
 retail store experience, 145-146, 247-248
 role of herbalists, 295
 school accreditation, 203-204
 shamanic drumming and journeying, 177
 spirituality, 259-260
 teaching herbalism, 219-222
 use of herbs daily, 305
Elecampane, 157, 288
Elements of Good Health class, 137-139
Elijio, Don, 92-93
Emerald Valley Herbs, 149
Emotional pain relief, 234
Empirical experience, 132-133
Endangered plants
 bioregional, 300-301
 black cohosh, 187, 281, 300, 302, 312
 complication of issues, 7-8
 cultivation of, 278-280
 goldenseal, 75, 106, 187-188, 300, 302, 312, 313
 Green Mountain Herbs, 106
 herbal industry role, 302-303
 impact of industry, 63-69
 impact of land development, 299-300
 listing, 333-334
 organic growing, 312-313
 role of manufacturing, 116-117
English herbalism practice, 284, 293
Environment
 education, 183-184
 relationship to organism, 13
Ephedra, 229
The Essential Book of Herbal Medicine, 36
Ethnobotanical concepts and education, 183-184
Ethnobotany, 214, 258-259
European herbalism, 240-242, 248, 278
Evaluation of herbs scientifically, 16-17
Exercise and health, 138-139
Expansion and potential of American herbalism, 209-210

Index

Facilitators of education and healing, 54
Fad medicine, 126
Fall/redemption philosophy, 84-85
False unicorn *(Chamaelirium luteum),* 116
Farmer's market movement, 147
FDA. *See* Food and Drug Administration
Federal Trade Commission (FTC), 107, 111
Female Formula product, 273
Feverfew, 142, 275
Findhorn, 152
Fish and wildlife government agency, 117
Flavonoids in milk thistle, 235-236
Flexnor Report, 208, 291
Flower essences, 166, 170-172, 219-220
Folklore
 herbalism and, 39-44, 190-191
 market setting, 281-282
 science and. *See* Science and folklore
Food Advisory Committee Working Group on Good Manufacturing Practices for Dietary Supplements, 273
Food and Drug Administration (FDA)
 DSHEA. *See* Dietary Supplement Health and Education Act
 Food Advisory Committee Working Group on Good Manufacturing Practices for Dietary Supplements, 273
 FTC versus, 107
 good manufacturing practices, 314
 language of science, 208
 licensure, 55
 marketplace therapeutics, 193
 purpose, 191, 201, 204-206
 regulation power, 261
Formulas for herbs, 220
Foster, Steven, 302
Fox, Matthew, 84-85
Franciscan community, 218
Frawley, David, 47

From Doctor to Healer, 12
FTC (Federal Trade Commission), 107, 111
Future of herbalism
 American Herbalists Guild, 251-252
 healing potential realized, 331
 Martin, Corinne, essay, 264-265
 Soule, Deb, essay, 306-307
 Trenkle, Peeka, essay, 324-325

Gagnon, Daniel, 27, 137-147
Gaia, 162
Garden Apprentice Program, 149
Garden project, 262-264
Gardening classes, 158-159
Gardner, Leslie, 48, 96, 149-160
Garlic, 15, 147
Gerard, John, 40
Germ theory, 303
German Commission E Monographs, 107, 288
Germany
 German Commission E Monographs, 107, 288
 influence on the FDA, 203, 205
 standardized extract, 242
Gilday, Kate
 at-risk flower essences, 170-172
 at-risk medicinals, 161-162
 background, 161
 bioregional use of plants, 77
 case study panel discussion, 58
 commerce of at-risk medicinals, 162
 commercialization, 42, 57
 communion with plants, 46
 disconnection from land, 166-167
 domino effect of at-risk medicinals, 168
 ecology, 66, 68, 75-76, 77
 endangered plants, 78-81
 ethical wildcrafting, 163-164
 herbs in the woodlands, 165-166
 knowing medicinal plants, 168-169
 mainstream herbs, 169-170

Gilday, Kate *(continued)*
    plant use education, 163
    plant-human relationship, 164
    religion and spirituality, 85-86, 87, 95
    whole individual healing, 27-28
Ginger, 244, 300
Ginkgo
    monopopularizing, 70, 71
    television advertising, 225
    uses of, 15, 275
Ginseng
    cultivated product, 65-66
    endangered plants, 300
    flower essences, 172
    long-term growth pattern, 80, 163
    market popularity, 63, 165
    monopopularizing, 70, 71, 119, 196
    organic growing, 7
    overharvesting, 302
    poaching, 118
Ginsengers, 162
Giving back, not only borrowing, 90-91
Gladstar, Rosemary
    background, 173-174, 276, 306
    classroom teaching, 158, 175-177, 179-181
    ecology, 72-73
    environmental and ethnobotanical concepts, 183-184
    experiential learning, 173-174
    herbal lifestyle, 21
    history of herbalism, 185-186, 200-201, 202, 206, 209
    judgmental people, 177-178
    new herbal consumers, 181-183
Gladstar Family Herbal, 173
GMH. *See* Green Mountain Herbs
GMPs (good manufacturing practices), 102-103, 273, 313, 314
Goal-setting for holistic health, 139
Gobo, 271
Gold thread, 75-76, 168
Goldenrod as allergy remedy, 108
Goldenseal
    berberine property, 75, 76
    CITES listing, 117-119

Goldenseal *(continued)*
    domino effect, 75, 77, 168
    endangered plants, 75, 106, 187-188, 300, 302, 312, 313
    habitat loss, 115
    market popularity, 63, 165
    misuse, 52
    monopopularizing, 196, 281
    organic growing, 7, 146, 147, 271, 278-280
    Purple Coneflower Herbals, 267, 270
    self-strangulation, 80, 163
    supply and demand, 64, 65
Good manufacturing practices (GMPs), 102-103, 273, 313, 314
Grandfathering of traditional information, 261-262, 282
Grassroots organization, 22-24, 201
Great Life, 237
Green, James
    aesthetics and spirituality, 11, 18
    background, 11-12, 185
    bioregionalism, 196-197
    classroom teaching, 151
    corporatization of herbalism, 189-190
    endangered plants, 187-188
    folkloric herbalism, 41
    gardening, 159
    herbal renaissance, 186
    history of American herbalism, 185-186
    holistic healing, 26, 191-192
    independence, 23
    individual power in healing, 192-193
    The Male Herbal, 22, 185
    marketplace therapeutics, 193-195
    medicine and healing, 188-189
    monopopularizing, 71, 73-74, 195-196
    nature and herbalists, 30
    science, 37-38, 39
    science and folklore, 187, 190-191
    technology of independence, 86
Green Hope Farm, 166
Green Medicine: An Eight Month Training in Western Herbalism, 319

Green Mountain Herbs (GMH)
  certifying efficacy, 106-107
  consumer expectations, 104-105
  educating consumers, 110-111
  endangered plants, 106
  federal regulations, 103-104
  marketing issues, 102-103
  mission, 105-106
  national sales, 100-102
  overview, 99-100
Green Nations Gathering, 41, 58, 123, 323
Green world connection to herbalism, 224-225
Grief and respiratory problems, 235
Group mind, 220
Growing At-Risk Medicinal Herbs, 113
Gutenberg's printing press, 296

Habitat reduction
  alien species, 301-302
  echinacea in Kansas, 116
  encroachment on environment, 161-162
  problem for at-risk plants, 113-115
Harvesting of at-risk plants
  care for plants during harvest, 161-162
  ethical wildcrafting, 163-164
  as livelihood, 68
  overharvesting, 271
  trowel and clippers method, 67
Hawthorn *(Crataegus oxycantha),* 40, 90, 108
Health, holistic, 137-139, 140-141
Health care
  alternative, 260-261
  American, 244, 246, 291-293
Heart-centered herbalism, 197, 218
Heisenberg uncertainty principle, 44, 45, 132
Henge of Keltria, 217
Herb Pharm, 113, 162, 206
The Herb Quarterly, 237

Herbal actions, 105
Herbal Defense, 237
Herbal gold rush, 86-87, 239
Herbal Healing for Women, 173
Herbal Medicine from the Heart of the Earth, 309
The Herbal Medicine Maker's Handbook, 185, 187
Herbal Product Industry (HPI)
  herbs versus herbalism, 276
  lifestyle choices, 274
  monetary profits, 18
Herbal renaissance. *See* American herbal renaissance
HerbalGram, 18
Herbalism
  biomedicine enhancement, 35-39
  defined, 30, 238, 283
  folklore and, 39-44
  green world connection, 224-225
  herbs versus, 207, 238-239, 276-277
  medical science and, 223-224, 256
Herbalism: Gaia in Action in Therapeutic Herbalism, 199
Herbalist
  defined, 255-256
  roles, 283-284, 293-295
Herbalist & Alchemist, 165, 206
HerbNET, 126
Herbs, Etc., 137
Herbs versus herbalism, 207, 238-239, 276-277
High-performance liquid chromtography (HPLC), 104
Hippies, 201-202
History of American herbalism, 185-186, 199-200
Hobbs, Christopher, 58, 202, 276
Hoffmann, David
  background, 199
  borrowing of American herbalism, 89
  changes in American herbalism, 204
  consumer safety, 51
  essence of American herbalism, 200-202
  expansion of herbalism, 209-210

Hoffmann, David *(continued)*
    FDA and medical establishment, 204-206
    herb industry, 206-207
    history of American herbalism, 199-200
    holistic herbalism, 22, 37, 211-212
    intellectual property rights, 214-215
    licensure, 212-214
    science and herbalism, 203-204, 208-209, 210-211
    scientific evaluation of herbs, 16
Holistic
    defined, 30, 191, 320
    health, 137-139, 140-141, 169-170
    paradigm, 12-13, 14, 35
Holistic healing
    allopathic medicine versus, 138, 269
    marketplace, 274-275
    nature of, 191-192, 273-274
Holistic herbalism
    allopathic medicine versus, 138, 143, 269, 289, 305-306, 319-320
    biomedicine and physics comparison, 35
    body-mind-spirit connection, 26-28, 192, 234-236
    defined, 12, 20, 30, 33
    Earth connection, 21-22
    grassroots organization independence, 22-24
    herbs as lifestyle, 20-21
    importance of place, 87-88
    individual attention, 28-31
    mainstream, 153, 317, 327, 329-330
    overview, 14, 330
    practicing of, 211-212
    science and folklore, 24-26
Homeless community garden project, 262-264
Hopman, Ellen Evert, 20-22, 62, 217-226
Horizon Herbs, 113
Hormone therapy
    black cohosh, 15
    red clover, 15

Horne, Steven, 24, 27, 40, 227-236
Horsetail, 74
Hospitals
    Center for Alternative Living Medicine, 283
    herbalist privileges, 245, 260-261
HPI. *See* Herbal Product Industry
HPLC (high-performance liquid chromtography), 104
Humanism and modern physicians, 39
Humanistic paradigm, 12-13, 14, 35
Human-plant relationship, 164
Hydrastis *(Hydrastis canadensis)*, 312. *See also* Goldenseal
Hydrocortisone effect of licorice, 40
Hyperforin, 77
Hypericin, 77, 142, 215
Hypericum perforatum. See St. John's wort

Independence of grassroots organization, 22-24
Individual attention in treatment, 28-31
Industry of herbs, 206-207
Inflammation, 147
Insurance
    acupuncturists, 247
    herbalist visits, 53-54, 244-245
    physician source of income in herbals, 293
Insure Herbal product, 273
Intellectual property rights, 214-215
International Herb Symposium, 41
Internet, usefulness of, 296-297
Intimacy of holistic herbalism, 28-31, 257-260
Isolationist attitude, 90

Japanese barberry, 168
Jayhawk Audubon Society, 61
Jewelweed, 120

Jocelyn Botanical Society, 300
Jones, Feather, 202
Judgmental opinions of herbalism, 177-178

Kava, 15, 275
Keltria, 217
Kether-One, 227
Khalsa, Karta Purkh Singh (K. P.), 52, 53-54, 86-87, 237-253
Khalsa, Siri Gian Singh, 93
Knowledge outstripping skill, 83

Labeling of products, 314-315
Language of science, 36-37, 208-209
Lasix, 143
Lavender, 226
Lay versus clinical herbalists, 124-126
Legalize adulthood, 244
Lemon balm, 77
Let's Live, 237
License to kill, 133-134
Licensure. *See also* Certification
  competency and, 230-232, 261, 284
  defined, 50
  difficulty in obtaining, 208-209
  overview, 49-50
  peer review, 212-214
  reasons for, 50-54, 144-145, 212-213, 244-245, 284-286
  reservations, 54-60, 193, 225-226, 243-247, 261, 284-286, 325-326
  science and, 297-298
Licorice, 40
Lifestyle of herbs, 20-21, 270
Listing of endangered plants, 333-334
Livelihood of harvest, 68, 162
Liver
  dandelion detoxifier, 297
  impact of anger, 235
The Lost Language of Plants, 61

Low Dog, Tieraona, 91
Lust, John, 200

Magic bullet
  corporatization of herbalism, 189
  media hype, 70
  monopopularizing, 71, 195-196
  one-step remedy, 187
  public desire for, 71-72
Mainstream holistic herbalism
  consciousness of concepts, 317, 327
  future of, 329-330
  Gardner, Leslie, essay, 153
  Gilday, Kate, essay, 169-170
  monopopularizing, 119-120
  paradigm shift, 296-297, 330
  professionalization, 53
Making Plant Medicine, 113
Male Formula product, 273
The Male Herbal, 22, 185
Manufacturing, impact on endangered plants, 116-117
MAO inhibitors, 287
Marketing of product, 102-103, 151-152, 162
Marketplace
  boom in herbal sales, 311-312
  therapeutics, 193-195
Marshmallow root, 64-65, 76, 304
Martin, Corinne, 19, 28, 36-37, 41, 255-265
Massage therapy, 246
Master herbalists, 157
Materia medica
  domino effect, 72
  education, 145, 220
  top ten herbal products, 72
Mayans
  autobiography of experience, 89
  belief of life, 48
  reciprocity of knowledge, 91
  spirituality, 92
McCleary, Annie, 42-43, 77, 94, 267-272

McGuffin, Michael
  background, 273
  consumer responsibility, 275-276
  earth connection, 18
  ecology, 63-66, 68, 69
  endangered plants, 278-280
  folklore in the market, 281-282
  herbs versus herbalism, 276-277
  holistic healing, 273-275
  monopopularizing, 72, 280-281
  self-responsibility of healing, 23
  use of herbal products, 277-278
McZand Herbals, 23, 63, 64, 273
Media hype of herbs, 71-73
Medical Herbalism, 199, 309
Medical Paradigm Continuum, 12-13
Medicine function in healing, 188-189, 193
Medicine power, 235
Medline, 205, 206
Melting pot syndrome, 85-86, 89
Memory, ginkgo for improvement of, 15
Menopausal Years: The Wise Woman Way, 143
Menopause
  black cohosh, 241
  Remifemin, 302
  Trillium essence, 172
Messengers of the Gods, 93
Midwifery, 262, 322
Milbradt, David, 36, 53, 88, 283-289
Milk thistle, 211, 235
Mills, Simon, 26, 29, 36, 286
Mind-body connection, 17-18, 26-28, 192
Modality of herbalism, 250-251
Moderation, holistic concept, 274
Monarch butterfly and BT corn, 303-304
Monetary profits, 18-19
Monopoly of science, 190
Monopopularizing
  complications of herbalism study, 71
  defined, 70
  echinacea, 242

Monopopularizing *(continued)*
  ecosystem threat, 70
  environmental education, 73-74
  limited mainstream materia medica, 119-120
  magic bullet, 71, 195-196
  media hype, 71-73, 280-281
Montgomery, Pam, 209
Moore, Michael, 137, 200, 214
Moratorium on endangered medicinals, 68
Mormon herbalism, 202
Motherwort, 77
Mullein, 74, 146
Murray, Michael, 214
Mushroom, 132-133
Myers, Norma, 185, 195

Narby, Jeremy, 46
Nasturtium seeds, 107
National Center for Complementary and Alternative Medicine, 62
National Institute of Medical Herbalists, 293
National Institutes of Health, 62
National School of Phytotherapy, 322
Native American
  medicine power, 235
  sweat lodge, 90, 91
Naturopathic physicians, 250
Nearing, Helen, 303
Nervine tonic (skullcap), 17
Netherlands health care, 244, 246
Network of herbalists, 41
The New Holistic Herbal, 199
New York Botanical Gardens, 91
News fasting, 143-144
Northrup, Christiane, 322
Nutrition and herbalism, 287

Oats, 77
One-on-one apprenticeship, 181
Open systems, 13, 19

Oregon grape, 75-76, 168
Organic foods, 141, 146-147
Organic growing of herbs
  echinacea, 147, 271
  economic investment, 65
  goldenseal, 146-147, 271, 278-280
  increased demand for, 146-147
  source of endangered plants, 7, 312-313
Organism relationship to environment, 13
Oriental medicine, 295-296. *See also* Chinese medicine
Osha, 77, 115, 271
Out of the Earth: The Essential Book of Herbal Medicine, 26, 286

Pacific NW Herbal Symposium, 309
Paracelsus, 219
Paradigms
  bioregional herbalism, 77
  bridging herbalism and medical science, 256, 257, 258, 264-265
  building on other paradigms, 19-20
  domino effect with endangered plants, 304
  holistic, 12-13, 14
  holistic versus allopathic use of herbs, 305-306
  humanistic, 12-13, 14
  licensure and science, 298
  mainstream shift, 296-297, 330
  technocratic, 12-13, 14
Parenting Series: A Holistic Approach to Raising Children, 319
Parthenolides, 142
Patient-clinician relationship, 129-131
Patterns of health, 149-150
Pauli exclusion principle, 132
PDR (Physician's Desk Reference), 204-205, 294
Pelechaty, Orest, 25-26, 56-57, 291-298

People of the Earth: The New Pagans Speak Out, 217
Peppermint tea, 241
Pharmaceuticals
  herbs versus, 15
  newcomer to herb business, 110
Pharmacognosy, 105
Pharmacopoeia realization of herbs, 204-206
Physician's Desk Reference (PDR), 204-205, 294
Phytomedicines, 16
Pills versus herbs, 139-140
Pipsissewa *(Chimaphila umbellata),* 66-67, 114
"Pissing in the wind" expression, 214
Place, importance of, 87-88
Plant allies, 156
Plantain, 188, 289
Plant-human relationship, 164
Planting the Future, 76-77, 173
Pleurisy root, 61, 300-301
Poaching, 118
Pollinators for plant survival, 299-300
Postpartum use of Trillium essence, 172
Prairie decimation, 61-62, 116
Prechtel, Martin, 48, 89, 95
Predictability of herbs, 236
Preserving herbalism, 153-154
Prevention, 18
Pritikin diet, 141
Probability calculations, 45
Professional Herbalist Certification Course, 237
Professionalization
  AHG vision, 251-252
  divisiveness of, 60
  future of, 57
  legitimization to the mainstream, 53
  licensure, 243-244. *See also* Licensure
  overview, 49-50
Properties of plants, 196
Property rights, intellectual, 214-215
Prostate issues and saw palmetto, 228
Proteins, 287

Prozac substitute of St. John's wort, 76
Pruning, 155-156
Prunus serotina (wild cherry bark), 90
Psychology and herbalism, 47
Purple Coneflower Herbals, 267
Purple loosestrife, 302
Pygeum *(Pygeum africanum)*, 312

Quantum mechanics, 45
Quantum physics, 34-35, 44, 47-48

Reciprocity
   herbalists and plants, 94, 96
   of traditions, 91
Red clover, 15
Red root, 147
Reductionist scientific thinking, 233-234
Regulations
   American health care, 244
   benefits, 242-243
   FDA. *See* Food and Drug Administration
   gray zone for GMP, 102-104
Relationship model, 45
Religion and spirituality, 84-85
Remifemin, 302
Renaissance. *See* American herbal renaissance
Respiratory congestion
   goldenseal, 75
   grief as cause, 235
Retreating of herbs into woodlands, 165-166
Rifts in herbal community, 326-327
Rituals, 154-156
Robbins, John, 305
Robinson, Hortense, 93
Romm, Aviva, 50
Roots
   goldenseal roots, 64, 75
   habitat loss, 117
   harvesting of, 51-52
   harvesting of endangered plants, 63-64

The Roots of Healing: A Woman's Book of Herbs, 299
Rumex crispus (yellow dock), 74, 271

Sacred Plant Medicine, 41-42
Safety. *See also* Licensure
   consumer education, 111
   drug interactions, 183
   herbal remedies, 228-229
   licensure, 59, 214, 232
   self-medication, 292
Sage Healing Way, 173
Sage Mountain, 173, 180
Sales statistics, 11-12, 18
Salmon, 118
Salmon, Enrique, 84, 93
Saponins, 208
Sassafras, 64, 279
Sastun, 91, 92
Saw palmetto
   monopopularizing, 70
   unpalatable taste, 211
   uses of, 228
Scalzo, Rick, 276
Schizandra, 147
School accreditation
   difficulty of naturopathic schools, 203-204
   grandfathered folk healers, 261-262
   licensure of graduates, 51-52
Science
   in American herbalism, 203-204, 210-211
   defined, 34
   enhancement of herbalism, 35-39
   evaluation of herbs, 16-17
   language of, 36-37, 208-209
   medicine and herbalism, 223-224, 256
   misgivings of herbalists, 37-38
   monopoly, 190
   overview, 33-35
   training for herbalists, 221-222

Science and folklore
  in evaluating herbs, 142
  Green, James, essay, 187
  imbalance, 159
  links, 44-48
  Milbradt, David, essay, 286-288
  reductionist, 233-234
  Tilgner, Sharol, essay, 315-317
  usefulness of, 24-26, 132-133
Scutellaria lateriflora, 17
Secrets of the Talking Jaguar, 48
Sedative action of cyanogenic glycosides, 90
Self-medication, 228-229, 292, 296
Self-responsibility for healing
  difficulties, 229-230
  herbalists as facilitators, 325
  holistic health, 23
  individual power, 192-193
  learning about the body, 23-24
  proper use of herbs, 277-278
Sensitization of self, 5-6
7song, 58, 283, 324-325
Shalimar Gardens, 309
Shamanic drumming and journeying, 177
Shamans sensitized to plants, 46
Sheehan, Molly, 166
Shelf-space fighting, 189
Sisters on a Journey, 8
Skullcap, 16-17
Sleep
  soporific herbs, 105
  valerian or California poppy, 37
Slippery elm
  monopopularizing, 196
  organic growing, 279
  replacement, 64-65, 76, 304
  supply and demand, 64
Smith, Ed, 158
Soporific (sleepiness), 105
Soule, Deb
  background, 299
  earth connection, 19, 303-304
  ecology, 67, 76, 77, 299-303, 304
  future of herbalism, 306-307

Soule, Deb *(continued)*
  Gardens at Avena, 303
  herbalism movement, 306
  holistic versus allopathic use of herbs, 305-306
  respect for plants, 81
  self-regulation of herbalists, 261
Spirituality
  in America, 84-85
  borrowing of American herbalism, 89-90
  connection to mind and body, 17-18, 26-28, 192
  herbal education, 259-260
  herbal healing, 322-323
  herbalist, 84, 88, 92-96
  infusion in daily life, 272
  infusion into work, 267-269
  in medicine making, 42-43
  overharvesting, 271
  plant communication, 150-151
  rituals, 154-155
Spontaneous Healing, 17
Springtime herbalist, 185
St. John, Gloria, 12, 13, 14, 19-20, 29
St. John's wort *(Hypericum perforatum)*
  addition to other products, 302
  alien species in habitat loss, 301
  clinical trials, 108
  complexity of, 157
  domino effect, 77
  efficacy certification, 108
  growing at home, 74
  hypericin, 142, 215
  MAO inhibitor, 287
  media hype, 230
  monopopularizing, 70, 71
  organic growing, 146
  Prozac substitute, 76
  television advertising, 225
  uses of, 15, 228
  wildcrafting, 271
Standardized herbs, 142, 298
Stapp, Henry, 44
Starting an herb business, 109-110

Statistics of use, 11-12
Stenn, Frederick, 39
Substantiation of claims, 282
Substitution of plants, 74-78, 168-169.
    See also Domino effect
Summertime herbalist, 185-186
Sundew, 165, 312
Supply and demand issues, 63-69
Sustainability, 258-259
System tonics, 36

Tao of Physics, 317
TCM (traditional Chinese medicine), 149, 213, 285-286
Teaching herbalism, 219-222
Technocratic paradigm, 12-13, 14, 35
Technology of independence, 22-23, 86, 185
Terra Nova Medicinal Plant Reserve, 91
Therapeutics in the marketplace, 193-195
Thin line chromatography (TLC), 313
Three-way relationship model, 45, 47
Tierra, Michael, 185-186, 200, 202
Tilgner, Sharol, 309-317
Time focus on top-selling herbs, 71
TLC (thin line chromatography), 313
Tom's of Maine
    company values, 101
    Green Mountain reformulations, 106
    homeless community garden project, 264
    newcomer to herb business, 110
    purchase of Green Mountain Herbs, 99, 100
    research department, 104
To-watch plants listing, 77, 334
TRAFFIC International, 117
Traffic North America, 281
TRAFFIC USA, 302
Transplanting, 155-156
Trauma care, 250
Tree vision, 330-332

Trenkle, Peeka
    background, 319
    diversity of herbalism, 58
    herbal community rifts, 326-327
    herbal education, 321-322
    herbalism as consumerism, 323-324
    holistic versus allopathic herbalism, 319-320
    licensure, 325-326
    mainstream herbalism, 327
    marketplace independence, 73
    professionalization, 54-55, 56
    specific herbs for conditions, 15
    spirituality in healing, 322-323
    vision of herbalism, 324-325
Trillium, 172, 300
Trotsky, 208
Tyler, Varro, 16

Umbilical cord syndrome, 149
Unconscious power, 4-6, 330, 331
Underworld, 55
United Plant Savers (UpS)
    at-risk criteria, 115-116
    at-risk plant listing, 333
    board member Michael McGuffin, 273
    board of directors, James Green, 185
    education of endangered plants, 117
    founder Rosemary Gladstar, 173
    gold thread as endangered, 75
    goldenseal listing on CITES, 117
    habitat loss, 115
    herbal industry monitoring, 302
    influence over companies, 69
    listing plants as endangered, 300-301
    Membership Conference speakers, 113
    Oregon grape as endangered, 75
    proactive approach, 120-121, 153
    resistance to moratorium on endangered medicinals, 68

United Plant Savers *(continued)*
   risk of medicinals for specific diseases, 77
   to-watch plant listing, 334
UpS. *See* United Plant Savers
U.S. Fish and Wildlife Service, 280

Valerian, 37, 299
Values
   American Herbalists Guild, 298
   communication to students, 220-221
   Tom's of Maine, 101
Variety in herbalism, 306
Village licensing system, 51
Vital force, 234, 268
Vitalistic herbal healing, 13, 27
von Bertalanffy, Ludwig, 13

Water needs, 139-140
The Web of Life, 94
Web sites
   Medline, 205
   United Plant Savers, 333
Weed, Susun, 143, 202, 270
Weil, Andrew, 17-18, 143-144, 322
Wesley, John, 202
Wetland development, 300
White House Commission on CAM, 50-51, 53, 56
Wild cherry bark *(Prunus serotina)*, 90
Wild gentian, 116

Wildcrafters
   gathering large quantities, 66, 163-164
   large company, 114
   livelihood of harvest, 68, 162
   proper practices, 271
   trowel and clippers harvesting, 67
Willow bark, 258
Winston, David, 232
Winter herbalist, 185
Wise Acres, 309
Wise Woman
   ecosystem preservation, 242
   education for herbalists, 157
   herbal healing approach, 13, 58
Wise Woman Herbals, Inc., 309-311, 314
Women's health, 319
Wood, Matthew, 55, 168, 234, 283
World Wildlife Fund, 302
Worsley, J. R., 285
Wounds, treatment with yarrow, 40, 233

Yarrow *(Achillea millefolium)*, 40, 90, 233
Yellow dock *(Rumex crispus)*, 74, 271
Yellow root, 75, 168
Yoga, 238
Your Body's Many Cries for Water, 139-140
Yudin, Sidney, 149

Zand, Janet, 273, 276

Order a copy of this book with this form or online at:
http://www.haworthpress.com/store/product.asp?sku=5224

HERBAL VOICES
American Herbalism Through the Words of American Herbalists

_____ in hardbound at $59.95 (ISBN: 0-7890-2203-6)

_____ in softbound at $34.95 (ISBN: 0-7890-2204-4)

Or order online and use special offer code HEC25 in the shopping cart.

COST OF BOOKS_____	☐ **BILL ME LATER:** (Bill-me option is good on US/Canada/Mexico orders only; not good to jobbers, wholesalers, or subscription agencies.)
POSTAGE & HANDLING_____ (US: $4.00 for first book & $1.50 for each additional book) (Outside US: $5.00 for first book & $2.00 for each additional book)	☐ Check here if billing address is different from shipping address and attach purchase order and billing address information. Signature_____
SUBTOTAL_____	☐ **PAYMENT ENCLOSED:** $_____
IN CANADA: ADD 7% GST_____	☐ **PLEASE CHARGE TO MY CREDIT CARD.**
STATE TAX_____ (NJ, NY, OH, MN, CA, IL, IN, & SD residents, add appropriate local sales tax)	☐ Visa ☐ MasterCard ☐ AmEx ☐ Discover ☐ Diner's Club ☐ Eurocard ☐ JCB Account # _____
FINAL TOTAL_____ (If paying in Canadian funds, convert using the current exchange rate, UNESCO coupons welcome)	Exp. Date_____ Signature_____

Prices in US dollars and subject to change without notice.

NAME_____
INSTITUTION_____
ADDRESS_____
CITY_____
STATE/ZIP_____
COUNTRY_____ COUNTY (NY residents only)_____
TEL_____ FAX_____
E-MAIL_____

May we use your e-mail address for confirmations and other types of information? ☐ Yes ☐ No We appreciate receiving your e-mail address and fax number. Haworth would like to e-mail or fax special discount offers to you, as a preferred customer. **We will never share, rent, or exchange your e-mail address or fax number.** We regard such actions as an invasion of your privacy.

Order From Your Local Bookstore or Directly From
The Haworth Press, Inc.
10 Alice Street, Binghamton, New York 13904-1580 • USA
TELEPHONE: 1-800-HAWORTH (1-800-429-6784) / Outside US/Canada: (607) 722-5857
FAX: 1-800-895-0582 / Outside US/Canada: (607) 771-0012
E-mailto: orders@haworthpress.com

For orders outside US and Canada, you may wish to order through your local sales representative, distributor, or bookseller.
For information, see http://haworthpress.com/distributors

(Discounts are available for individual orders in US and Canada only, not booksellers/distributors.)

PLEASE PHOTOCOPY THIS FORM FOR YOUR PERSONAL USE.
http://www.HaworthPress.com   BOF04